Algorithms for Construction of Reality in Physics
Volume 2

Schrödinger's Cat Smile

Authored by

Sergey P. Suprun
Semiconductor Physics, Siberian Branch of the Russian Academy of Sciences, Novosibirsk, Russia

Anatoly P. Suprun
Psychology Department, Lomonosov Moscow State University, Moscow, Russia

&

Victor F. Petrenko
Psychology Department, Lomonosov Moscow State University, Moscow, Russia

Algorithms for Construction of Reality in Physics

Volume # 2

Schrödinger's Cat Smile

Authors: Sergey Suprun, Anatoly Suprun & Victor Petrenko

ISSN (Online): 2212-8514

ISSN (Print): 2589-3572

ISBN (Online): 978-981-5049-66-4

ISBN (Print): 978-981-5049-67-1

ISBN (Paperback): 978-981-5049-68-8

need for a court order if at any point you breach any terms of this License Agreement. In no event will any delay or failure by Bentham Science Publishers in enforcing your compliance with this License Agreement constitute a waiver of any of its rights.

3. You acknowledge that you have read this License Agreement, and agree to be bound by its terms and conditions. To the extent that any other terms and conditions presented on any website of Bentham Science Publishers conflict with, or are inconsistent with, the terms and conditions set out in this License Agreement, you acknowledge that the terms and conditions set out in this License Agreement shall prevail.

Bentham Science Publishers Pte. Ltd.
80 Robinson Road #02-00
Singapore 068898
Singapore
Email: subscriptions@benthamscience.net

CONTENTS

FOREWORD 1

Currently, not only psychologists and philosophers are interested in the problem of consciousness, but also physicists, who conducted a number of experiments in the field of quantum physics (teleportation of quantum states, "erasing" of the past in the experiments with quantum eraser, and others) that have revealed a strange dependence of physical reality on the presence of an observer in it. Moreover, several attempts have been made to create the quantum theory that includes the observer's consciousness since this is the key factor in the mysterious phenomenon of wave function collapse, which according to Niels Bohr gave rise to "external reality". The work on quantum computers revived the interest in old unsolved problems in physics as well as debates of both physicists and philosophers.

Among the difficulties associated with the discussion of this interdisciplinary problem is the difference between the paradigms for considering physical and psychological realities. This book, written by a physicist and psychologists, attempts to resolve the problem of agreeing on these paradigms using a systems approach. Two types of reality representations are analyzed, namely, in terms of the classical object-based spacetime model in the system of consciousness and a spectral model in Hilbert's space, characteristic of quantum mechanics, in the system of the unconscious.

It is demonstrated that the psychosemantic approach makes it possible to derive relativistic laws of energy and momentum conservation in a semantic form beyond the physical paradigm and to expand the concept of the classical observer's reference frame to the system of observation that includes individual characteristics of the subject's "perception channel", which makes it possible to leave the boundaries of spacetime representation for the quantum region (beyond the boundaries of consciousness).

The authors discuss the possibility to construct both quantum and psychological theories of the unconscious in a new paradigm. It is important to note that the first attempt to use this approach dates back to Carl Jung, a known psychologist, and Wolfgang Pauli, a physicist, and a Nobel Prize winner.

The book is addressed to a wide range of readers interested in the modern problems in physics and psychology as well as the students and postgraduates who specialized in these areas.

Dmitry Ushakov
Academician of the Russian Academy of Sciences,
Institute of Psychology, Russian Academy of Sciences,
Moscow, Russia

FOREWORD 2

Rapid development of technologies over the last century has completely changed the face of our civilization. Just recently, it was most difficult to imagine the experiments in quantum physics, such as the teleportation of quantum states even with a velocity exceeding the speed of light or "erasing" of the past in the experiments with a quantum eraser. Here, it is high time to recall Niels Bohr and his statement that "Anyone who is not shocked by quantum theory has not understood it." The impression emerges that our concepts of reality and the very reality are separated by an abyss and this abyss widens with the advance in science. Truly crazy theories, unverifiable in experiments, appear in physics; perhaps, Bohr was right to say that "Your theory is crazy, but it's not crazy enough to be true." Most likely, a fundamentally different approach is necessary to finally gain an understanding of physics. It is appropriate to ask the question how our concepts of reality are actually formed if they are so far from the reality itself?

This monograph, prepared in collaboration with a physicist and psychologist, attempts to answer this question. Unlike other studies, the focus here is not on the criticism of the existing situation with the interpretation of the experiments in quantum physics but rather on the search for a way out. Using the methods of psychosemantics, the worldview of an individual is successively analyzed, which suggests an unexpected conclusion that man actually exists in the model of reality that is constructed by his unconscious. This model is object-based in its content and the objects are arranged in spacetime, emerging to be mental constructs, as was noted by Henri Poincare. An object-based space of qualities is constructed and used to deduce the laws of conservation of energy and momentum in a relativistic form even without the use of the hypothesis of the existence of physical spacetime.

Certainly, several assertions of the authors are rather bold, first and foremost, because it is yet unclear which of our long-held beliefs we are ready to abandon in order to accept reality. However, this attempt has the right to exist even by the mere fact that we have currently no other solutions. Presumably, only doubting common sense will allow us to find the truth.

This book is recommended to a wide range of experts, students, and postgraduates in both natural and social sciences and to all who are interested in the current problems in science.

Igor Neizvestnyi
Corresponding member of the Russian Academy of Sciences,
Institute of Semiconductor Physics, Siberian Branch, Russian Academy of Sciences,
Novosibirsk, Russia

PREFACE

The second volume of the series of monographs titled Algorithms for Construction of Reality in Physics continues the successive analysis of the form and content of the worldview perceived by humans. The fact that we are "submerged" in the model of reality that is constructed by our unconscious, that is, is not controlled by consciousness, is substantiated. As early as 100 years ago, Henri Poincare paid attention to how and why, for example, the perception of space could emerge although, at that time, this was a flash of a genius unsupported by any psychophysiological research data. Strange as it may seem, the problem of reality has become extremely relevant with the development of quantum physics. The experiments in this scientific area have illustratively demonstrated that our naïve notions about the objects residing in spacetime fail to fit the reality. Quantum teleportation, i.e., the transfer of quantum state at any velocity including that faster than the speed of light, demonstrates the absence of locality (or separability), which means that the integrity of reality has no spatial limits. The experiments with delayed selection of a "quantum eraser" type make it possible to change "the past", suggesting the absence of any time constraints. Thus, the "instrumentality" of their studies (the answer is numeric) brought physicists at the cutting edge of the research into reality rather than its model, suggested to us by our unconscious.

In the case when it is difficult to separate reality and its model, it is reasonable to study the mechanisms of how the model was constructed. A psychosemantic approach makes it possible to analyze the specific features of such construct and to infer what in it is God-given and what is evil. From this standpoint, it is of interest to consider an object-based space with qualities as unit vectors. This helps to answer the question of what are the principles of conservation of, for example, energy and momentum. We regard these principles as the laws of Nature rather that the rules according to which the model reality functions and the main requirements of which are logic and preservation of the content. As it happens, these "laws of Nature" in terms of relativity are deducible even without the hypothesis of spacetime existence.

In the context of this approach, it is reasonable to take a fresh look at the problems in quantum physics. With this in mind, it becomes clear that the attitudes of an object-based model of reality in our consciousness are the particular factor that prevents us from an open-minded consideration of the experimental results in this area of knowledge. We "see" objects residing in spacetime where they do not exist and have never existed. It is believed a priori that the so-called entangled pair is a pair of objects; however, in this case, they must "behave" as objects, have the properties of objects, and evolve in a spacetime frame. However, experiments illustratively demonstrate that this is not the case. Perhaps, they are not objects?

In this sense, it is also interesting to consider some particular problems associated with our consciousness that are suspiciously analogous to certain phenomena in physics. Here, we do not make any far-reaching conclusions; our goal was to merely attract attention to these analogies. The parallels between the oriental philosophy of Buddhism and modern scientific concepts have long been discussed in the relevant literature and this is not at all accidental. The western methodology has successively implemented mainly object-based decomposition of the world (completely free from the subject, the "apex of creation") and is continuing to develop it even in the systemic paradigm of quantum physics, thereby giving birth to the "centaurs", such as wave mechanics. The oriental scholastics from the very beginning developed a holistic, systems-based view on reality with the man as its inherent subsystem. Since all evolving subsystems are open and, correspondingly, linked to the system of individual consciousness, this made it possible to embrace other types of "consciousnesses" with different forms of reality representation and other properties inaccessible in the boundaries of our type of perception. They were the first to study the categories of integrity (emergence), purposefulness, the hierarchy of open systems (theory of the dharmas, Samsara and Nirvana, and so on) and analyzed the limitations of our language and thinking in the understanding of reality. Zen Buddhism has implemented an original psychotechnique allowing its adepts to trace the boundaries of their own consciousness with the help of specific limit questions, a kind of antinomies, the koans.

All these problems although in another form have again become relevant now. We have become captive to one of a multitude of forms of representation (modeling) of reality characteristic of our rather limited type of perception.

The issues described in the book and its content are interdisciplinary research, which most likely make the understanding not that easy. However, we believe that this will motivate the readers to search the literature by themselves for the facts that confirm or, perhaps, refute the described ideas. We will be glad to receive any sound criticism and are always open to discussion of any relevant issues.

CONSENT FOR PUBLICATION

Not applicable.

CONFLICT OF INTEREST

The authors declare no conflict of interest, financial or otherwise.

ACKNOWLEDGEMENTS

Many people have helped us during the work on this monograph and we would like to thank them for their assistance and support as well as for the time they spent reading the manuscript and their helpful advice.

The authors thank Viktor Ovsyuk, Yaroslav Bazaikin, and Vladimir Shumskiy for their discussion on the manuscript and helpful criticism.

Our special appreciation to our wives, both Galinas, for correcting the text and for their patience and understanding during our work on the manuscript.

We also acknowledge the work of our translator Galina Chirikova, who had to deal with the terminology of sciences so distant from one another and are especially grateful for her critical remarks, which we regard as very helpful for making the text clear.

This statement is to certify that all authors have seen and approved the manuscript being submitted. We warrant that the book is the Authors' original work. We warrant that the book has not received prior publication and is not under consideration for publication elsewhere.

Sergey P. Suprun
Laboratory of Heterostructure Physics and Technology
Institute of Semiconductor Physics
Siberian Branch, Russian Academy of Sciences
Novosibirsk, Russia

Anatoly P. Suprun
Laboratory of Psychology of Communication and Psychosemantics
Psychology Department, Lomonosov Moscow State University
Moscow
Russia

Victor F. Petrenko
Laboratory of Psychology of Communication and Psychosemantics
Psychology Department, Lomonosov Moscow State University
Moscow
Russia

INTRODUCTION

Freedom is the ability to think without prejudices and attitudes.
Midnight thoughts

This book continues and further develops the ideas detailed in the first volume, titled *Computers: Classical, Quantum and Others,* which makes it possible to revisit some of our views on reality. This is possible if we adopt the standpoint that we actually always deal only with a "subjective internal" model of reality constructed by our unconscious utilizing an evolutionarily established set of certain algorithms rather than with an "external objective" reality. The current advances in science, in particular, physics and psychology, provide a sufficiently comprehensive and convincing evidence for this statement [1]. The very life emerges to be the result of efforts of the manipulator, such as our unconscious, which always stays in the shadow. Moreover, the attempt to gain the insight into our own behavioral programs is a fascinating pursuit, which may enhance the resolution of many serious problems. This approach looks promising since it gives the possibility to take a fresh look at many things beyond physics per se. This also creates a universal basis for explanation of the phenomena belonging to different fields of knowledge just because of the mere fact that they are based on our perception of reality, first and foremost, at an unconscious level.

The heated debates of the beginning of the last century, which involved many outstanding physicists, including Niels Bohr, Albert Einstein, Erwin Schrödinger, Werner Heisenberg, Wolfgang Pauli, and David Bohm, had long died off. At that particular time, the theory of relativity and quantum mechanics were establishing; however, the debates were focused not on these novelties but rather on the Reality we were living in and our role in this Reality. Bohr and Einstein were the ideologists of two approaches to understanding reality. Although the nominal winner in this opposition was Bohr, who proposed the so-called Copenhagen interpretation of quantum mechanics based on the complementarity principle, the great many both then and now regard it as a tradeoff rather than the final resolution of the problem. The cozy Cartesian world collapsed with the advent of the principle of parallelism or, actually, the principle of "noninterference" of consciousness with the laws of fundamental physics, as was uttered by John von Neumann in his discussion of the phenomenon of wave function collapse [2]. Bohr believed that the world emerges from nonexistence at the very moment when a subject perceives it,[1] which is actually confirmed by the quantum-eraser experiments [3] interpreted in terms of the object-based paradigm. However, he did not consider the variant that it was the object-based representation of reality that emerged rather than the reality itself.

Later, John A. Wheeler, an American physicist, put down this view as "no phenomenon is a physical phenomenon until it is an observed phenomenon" and Pascual Jordan brought the Copenhagen denial of the observer-independent reality to its logical conclusion, claiming that "we ourselves produce the results of measurement" [4]. Actually, here we encounter a doublespeak: undoubtedly, we have an intuitive model of reality and the result is regarded as explainable if it meets this model. However, if we regard any fact or phenomenon that fits within the frame of the model as real, this turns everything upside down. That is why all these assertions may contain many evident contradictions: on the one hand, the nonexistence of quantum world as an objective reality before measurement is postulated and, on the other hand, its objective description is admitted (*i.e.*, an objective description of something that does not exist). Thus, the only reality is the consciousness of a subject since this particular consciousness is both the cause that brings the reality into being and the form of its representation. However, the mathematical tools of quantum mechanics lack any observer. The theory says nothing about the wave function collapse, a sudden jump in the state of a quantum system during measurements when a particular possibility becomes a "classical" reality. Note that the Copenhagen interpretation requires for an actual existence of the Universe that a perceiving observer be beyond this Universe. Otherwise, it could never reveal itself as Reality and would always remain in the state of superposition of many possibilities. Thus, even the "substance" as the basis of our world appeared to be a remnant of the worldview on the divine clay. Einstein totally disagreed with Bohr: "Do you really believe that the moon isn't there when nobody looks?" he asked Abraham Pais.

Thus, the parallel world "beyond perceptions", as an endless coffer full of things "reflected" in our consciousness, appeared to be a naïve metaphor of ancient philosophers. The subject/observer, who objectively "observed" the Universum as if being a "Holy Spirit", turned out to be inadequate to the new physics. The principle of objectivity demanded that the subject was excluded from the physical Reality since only objects with no free will had the right to be represented in physical theories. However, the subject although illegally and frequently in an indirect manner is still present in theories at least to define the frame of reference. Moreover, when defining the subject *as a certain unity,* it would be natural to believe that the subject must always be in the one and *only one frame of reference,* which is its own frame of reference!

Henri Poincare pointed that the physical phenomena taking place in different inertial systems and described in terms of proper metrics *are fundamentally incomparable,* which Einstein did. This is determined by the fact that the Lorentz transformations not only provide the conversion from one inertial system to another, but also automatically convert their spacetime metrics into each other [5]. This distinguished the Lorentz transformations from the Galilean transformations, which preserve the spacetime metric. Thus, the *invariance* of natural laws *relative to the Lorentz transformations* does not mean that the phenomena they describe proceed *identically* in different inertial systems, as Einstein believed. The Poincare–Lorentz relativity principle is implemented based on the *similarity* of kinematic relations rather than their *identity* (according to Einstein); correspondingly, *the differences* in the courses of physical processes in different reference frames *do not violate their equivalence.* The subjects are equivalent but do not perceive the Reality in an identical manner. In other words, they "live" in similar worlds. A real process considered by different observers is described within their own models and these models are similar rather than identical.

According to Poincare, we cannot experimentally verify the hypothesis on the isotropy of space. That is why the assertion on the constancy of the speed of light in the forward and reverse directions is the subject of agreement. Hence, any process of synchronization of spatially separated watches is the matter of convention *even within the same reference frame.* All this is a direct consequence of the prohibition, *i.e.,* the infeasibility to determine the absolute value of the speed of an inertial reference frame.

In the context of the so-called firewall paradox, which is the situation when two observers encounter conflicting descriptions of the same phenomenon, "a strong generalized complementary principle" is considered. More specifically, this means that all descriptions are confined not only to different spacetime regions, but also to the reference frames of different individual observers. Actually, this means that each individual observer has the own individual universe and the own event horizon. We only agree different elements of individual "mental maps" expressible in the second signal (sign) system of communication. For example, any possible implementation, *i.e.*, measurement, of a quantum system is a unique event that is perceived by different observers but in the own individual observer's reference frame. However, any interference between them is unfeasible owing to their distinguishability because the process of perception is individual. On the other hand, all possible measurements of a quantum system in all possible systems of reference do exist from the standpoint of Subject. In this sense, there is no difference between the superposition of states of a quantum system and all its possible implementations before the event of observation. Formally, both are describable in terms of the amplitude of probability as an entangled state of the quantum system or an arbitrary observer in the Subject's reference frame, which is unfortunately inaccessible to us since we are subsystems of the Subject. Although this representation is admissible and is frequently used, for example, in descriptions of the experiments with Schrödinger's cat (with the cat as an observer), it does not belong to our reality. This is a purely hypothetical view of the world unverifiable in our frame of reference because of the specific features in our perception of reality: we are unable to "see" the superposition state of a quantum system although this state does really exist. This situation is close in its meaning to the many-worlds hypothesis by Everett and Wheeler, which can be regarded as true but relative to the Subject rather than to us, which makes it useless. A similar situation is observable in psychology as well [5].

It is quite natural to assume that the *mental (purely utilitarian) image of the world,* created as early as our ancestors during the evolution and allowing them to cope with the "external reality" and adapt to it with their modest set of physical concepts about the outside world, is the initial foundation of our knowledge in any area. It is evident that this "mental visualization" of the information about reality available to us is based on certain axioms that are by far implicitly included into our theories and scientific concepts. For example, our consciousness partitions the visual world into objects and places them into a 3D space with a Euclidean metric. The encounter with the velocities beyond the range natural for our biology forced physicists to reconsider this thesis and propose a different metric in the special theory of relativity and even a different dimensionality. According to the view of Poincare, these "novelties" are dictated by convenience and simplicity of the theory rather than the reality itself, which does not give us unambiguous clues on this point. In his works, Poincare emphasized that the visible reality was only a projection of the visible world onto four-dimensional spacetime continuum. He believed that all our models of reality to a considerable degree rest on some incompletely comprehended conventions and are first and foremost determined by the goals (demands and motives) that were necessary to survive at the early stages of biological evolution.

Poincare was sure that any experiment could be adequately described and explained in many ways (theories). Selection of a particular model from the set of possible ones is rather arbitrary and is determined by the demand for simplicity and usability. According to Poincare, different groups of transformations can be ascribed to either "external" space or "internal" changes. For example, perspective (linear fractional) transformations in a certain manner "distort" the reality: it seems to us that objects decrease in size with an increase in the distance to them. However, we assume that this is a specific *objective* feature of our visual perception rather than an *objective* law of the physical space. As for other changes in our sensations, we relate them with our "internal" states, for example, the sensation of hunger. On the other hand, the Lorenz transformations [6] can be derived from the patterns of our *subjective perception* but physicists for some reason relate them to *objective* spacetime changes unlike the linear fractional transformations.

The special theory of relativity geometrized the united spacetime continuum. Naturally, new sensory illusions emerge in this new space. Time is inseparably connected with movement and movement, with force (or field)—the metaphors of the "threads" that sew together the "independent" and self-sufficient Platonic objects into the united cloth of the Universum. According to Roger Penrose [7], "there is nothing in the physicists' space–time descriptions that singles out 'time' as something that 'flows'." Time flows exclusively in our consciousness! This is worth thinking over.

When considering one's own past, an individual matches it against the own present. However, even the simplest linguistic analysis demonstrates that we initially understand the present as a certain time interval when a certain action is implemented *rather than a moment* (a point on the time axis, as is common in physics). For example, we say "I am writing (serving, lunching, etc.)" meaning not a time point but rather a time interval *when the action has not been completed.* In classical physics, this interval was illegally reduced to a point. This broke the axiomatics underlying the construction of physical and mathematical worldview initially natural for the human psychology since sensations cannot exist in a single moment even in terms of classical physics. For example, we need the time interval of at least one oscillation to hear an audio tone. Note that the existence of memory is already necessary at this very stage since the information obtained over this period should be somehow kept in mind. Consequently, the world deprived of memory has no time at all and thus *the beginning of time and the beginning of memory is one and the same.* From this standpoint, bringing back an intervalwise time estimation in quantum mechanics looks quite logical.

The EPR phenomena and Bell's theorem destroyed our concept of space and time, locality and causality. However, physicists hang on for the Platonic object representing a "perceivable" basis of the world as if catching the last straw. The overall physics, both classical and quantum, is constructed on an object-based metaphor despite that we had long ago left the boundaries of its applicability. Logical paradoxes of an object-based interpretation have led to the situation when physical theories have reduced to a set of recipes for computations in which the proper *physical* concepts and questions have become irrelevant ("Shut up and compute!" is the universal reply of the modern physics).

The absence of demarcation between perceptual and functional spaces and time leads to misunderstanding. All this takes place despite that Bertrand Russell as early as 1912 distinctly distinguished between the real and perceptual spaces. In particular, he wrote, "It is not only colours and sounds and so on that are absent from the scientific world of matter, but also *space* as we get it through sight or touch. It is essential to science that its matter should be in *a* space, but the space in which it is cannot be exactly the space we see or feel. <...> But this real shape, which is what concerns science, must be in a real space, not the same as anybody's *apparent* space" [8]. This was best expressed by Vivekananda in his metaphor: "Time, space, and causation are like the glass through which the Absolute is seen… in the Absolute there is neither time, space, nor causation" [9]. Interestingly, many physicists believe that oriental philosophy, which first and foremost focuses on the subject and his/her inner world, better complies with quantum mechanics as compared to outwardly oriented western philosophy, mainly focused on the object.

Most likely, this is the high time to look back to the origin of our concepts of the world and recognize the foundation on which our scientific knowledge is constructed. Evidently, we implement in our theories what is represented in our consciousness by perception, which is the only communication channel connecting us with the entity that we refer to as "external" reality. Actually, consciousness is the system of certain evolutionarily formed way of reality representation. In a psychological paradigm, that what is not the consciousness is beyond the boundary of consciousness, *i.e.*, unconscious, whereas in the physical paradigm, this is Reality.

Undoubtedly, Wolfgang Pauli and Carl Jung intuitively felt this link when they formulated the theory of collective unconscious and synchronicity (an analog of quantum teleportation in physics), where Pauli hoped to find the link between consciousness and wave function collapse. However, contrary to Occam's razor (not multiplying entities without necessity), he expected to find this link via resolution of a psychophysical problem, namely, the association between psychic (actually unconscious) and hypothetical Platonic "transcendent" world or the world "beyond sensations and perceptions", which he identified with the physical reality. Note that he assumed that the mechanisms that transform the excitations determined by physical world into conscious experiences of the psychic world are able to resolve the paradoxes of wave mechanics. However, these two worlds can interact *either physically or psychically*. In the former case, the psychic world reduces to physical and, as Neumann noted, the wave function does not collapse because of quantum entanglement. In the latter case, the imaginary "external world" becomes the unnecessary entity.

A systems approach to the problem makes it possible to escape many unnecessary metaphors and entities. Evidently, there are numerous ways to represent reality. An object represented in a physical space is not the same as in its spectral representation. The functional (spectral) Hilbert space lacks current physical time, which emerges only after a Fourier transform in a spacetime representation of the system of individual consciousness. Perception does not reflect the "external" world but rather translate the reality from the unconscious by transforming it with the help of psychic subsystems (unconscious) into an object-based spacetime form. Actually, the subject as if "browses" the Reality through a certain spectral window. Multiplication of the known relation for spectral window $\Delta\nu\cdot\Delta t\geq1$ by Planck's constant $\hbar$, easily transforms it into the indeterminacy relation $\Delta E\cdot\Delta t\geq\hbar$.

Quantum theory was constructed without any reliance on the clear understanding of its fundamentals, as a rule intuitively, using analogies and allusions to classical physics despite a great difference between these concepts. Presumably, the "birth traumas" during the emergence of this new theory still haunt it.

As is known, Schrödinger did not derive his famous equation describing, as he believed, certain "matter waves" but rather constructed this equation based on the de Broglie formula relating the "wavelength" and momentum of a particle. Schrödinger was inspired by a "noble goal" to save the new theory from "these damned quantum jumps", that were present in and, which was the most important, what actually oscillated. Schrödinger himself believed that these matter waves were as real as the other spatial types of waves. However, once the ψ function of electron in the hydrogen atom could be somehow interpreted as a three-dimensional wave, the wave function of two electrons in the helium atom had to be considered in a kind of obscure six-dimensional space. Later, Max Born proposed a ψ function interpretation using the concept of probability; thus, the wave function ceased to be a physical reality leaving for a mystical world of probabilities where it existed since that time. Quite soon, Bohr concluded that a "quantum object" did not exist anywhere at all until being observed and that the wave function would collapse to one of the possible states and the object would "materialize" in spacetime only after an event of observation (measurement).

Although Einstein looked towards returning to the reality concept of classical physics and tried to dispute the views of Bohr on quantum reality, it seems as if he also had some doubts on this point. "The more aristocratic illusion concerning the unlimited penetrative power of thought has as its counterpart, the more plebeian illusion of **naïve realism**, according to which things 'are' as they are perceived by us through our senses. This illusion dominates the daily life of men and of animals; it is also the point of departure in all of the sciences, especially of the natural sciences" [10].

Norbert Wiener gave the best description for this situation: "Physics is at present a mass of partial theories which no man has yet been able to render truly and clearly consistent. It has been well said that the modern physicist is a quantum theorist on Monday, Wednesday, and Friday and a student of gravitational relativity theory on Tuesday, Thursday, and Saturday. On Sunday he is praying ... that someone will find the reconciliation between the two views" [11].

An attempt of an object-based reality representation in a Hilbert space gave birth to weird "centaurs" of a "wave–particle" type. Our prejudices that reality must be represented only in a classical object-based form (*i.e.*, only as being perceived by our senses and in the form they are represented in our consciousness) have led to fruitless attempts to interbreed two fundamentally different systems, classical and quantum ones. The evolution of our scientific views has come a long way from the model of reality with a universal subject by Galileo–Newton to a multisubject (which is actually nonsense) model by Poincare–Lorentz and, eventually, to admission of the fact that each of us resides in the own individual reference frame determined not only by specific spacetime features but rather by individual specific features in the perception of reality.

The scientific understanding of the world is inevitably realized in semantically closed finite systems, which in physics correspond to closed (isolated) systems. Only such systems allow for motion integrals (conservation laws), reflecting the fact that the mathematically described structures in the absence of external impact should preserve certain semantic invariants of their initial formal description [6]. Evidently, Evolution as a process is irreducible to a change in the form at a given content (finite transformations) but rather changes the very content (or axiomatics of the theory), thereby giving birth to *new systems* that demand some other mathematics for their description. Although Evolution is implementable only in an open system, we are forced to again mathematically describe each of its stages (subsystem) as a *new finite system*. This *systemic decomposition* of reality necessarily raises the question on the "interaction" of these systems, that is, on the mechanisms underlying the translation of the "interaction" of these systems, that is, on the mechanisms underlying the translation of their content and the forms of its representation. It is evident that the notions of one system cannot be purely mechanically transferred to the other system (as it was done in the case of intuitive analog-based construction of quantum mechanics) without the relevant reinterpretation. Thus, we have to relate the observer's reference frame in a quantum system not to the spacetime characteristics, which are just absent in a Hilbert description, but rather to the system of translation, that is, individual perception characteristics of the "observer". It is also necessary to carefully consider the transfer of other classical notions to the new system (for example, vector and scalar values, since they will acquire different meaning there).

As it has emerged, the term "observer" in the modern science is ambiguous and controversial, differing in its meaning in the relativity theory and quantum mechanics. In the former, observer merely "describes" a considered process in a certain inertial reference frame from a position as if "above this reference frame"; however, in the latter, observer records the result of measurement, thereby determining the final state of a quantum system, which actually means the *interaction* with this system. Even if we reduce this interaction to the mechanisms and channels of translation (the transformation from one form of representation to another), we nevertheless have to take into account the *individual characteristics* of these mechanisms.

Once, Bohm noted that "...yet the fact that a great deal of what we see is ordered and organized in a form determined *by the functioning of our own bodies and nervous systems*[2] has very far reaching implications for the study of new domains of experience, whether in the field of immediate perception itself or in science (which generally *depends on instrumentally aided perception*, in order to reach new domains)" [12]. In other words, Evolution had created for us the corresponding mechanisms allowing for a spacetime reality representation, which, first and foremost, suited "biological survival" of our kindred at a macrolevel rather than at a microlevel. A mental map of reality available to our perception is represented in our consciousness; moreover, this map is "ordered and organized" in accordance with these mechanisms. Whereas the perception in the classical theory relates observer to the situation *within the frame of consciousness* (in a mental map), the perception in a quantum paradigm compares different forms of reality representation in different systems. This is the fundamental distinction of observer's reference frames in the relativity theory and quantum mechanics. **Thus, the reference frame of a subject** (observer's reference frame) **must comprise both the observation conditions (setting of experiment, devices, data processing, and so on) and the characteristics of the observer's perception machinery associated with the channel of content translation between the systems,** *i.e.*, **it must be actually considered as the** *frame of observation.*

Note that this expansion of the concept of the reference frame of an individual observer (hereinafter, observer's reference frame) makes it possible to adequately reconsider the experiments towards many mysterious phenomena of quantum mechanics, for example, wave function collapse during observation, *i.e.*, the process whereby an observation result appears on the mental map in the consciousness of a subject.

In our studies, we have approached the very "boundary of our consciousness", *i.e.*, the other system of reality representation. Evidently, some missteps in the interpretation of these representations are also likely in this situation. For example, a coordinate in Hilbert's space is representable in consciousness as a harmonic (a wave with a fixed frequency ν and a infinite in space and time. According to the de Broglie equation, we relate the particle's momentum to λ; correspondingly, the particle appears to be "smeared" over the entire space and behaves as a wave. On the contrary, a limited portion of the space whereto the particle can be localized has an infinite expansion in ν and, consequently, has an indefinite momentum.

The violation of causality in the EPR phenomena may be explained by that cause and consequence in the connected systems, namely, the consciousness in terms of the classical description (commutative algebra) and beyond it (noncommutative algebra), appear to be spaced apart and reside in different systems, which generates the illusion of cause-and-effect breakdown when they are jointly considered. For instance, the attempt to objectify entangled states in quantum mechanics brings about not a single object but rather a set of "objects" connected in an acausal, aspatial, and atemporal manner, which contradicts the relativity theory.

The world of consciousness is the world of classical physics. In terms of this approach, the Jung–Pauli hypothesis on the collective unconscious makes sense. The reality in this model can be considered as an infinite open evolving system wherein the finite subsystems that represent a certain aspect of its content emerge. In fact, the emergence of a new system (or its subsystem) means the emergence of a new Subject, which ranges from the Universe to an individual. What looks as the Big Bang in terms of a spacetime representation is actually the beginning of translation. Presumably, such object-based "visualization" of reality in our system of the Universum was evolutionarily beneficial for something since any system should be goal-oriented by definition [13].

We believe that the ideas highlighted in this book are able to resolve numerous paradoxes in physics and eliminate several questions at all. As we will show later, our intuitive model of reality is object-based and its inseparable attribute is the spacetime in which objects are naturally housed and their characteristics are described. Thus, a model that "works" well in the case of fulfillment of certain obligatory conditions can be object-based rather than the reality [14, 15]. What is more, a large part of the surrounding world not only belonging to the area of quantum physics but also, for example, living organisms, social phenomena, and many others appear to be describable using the system-based model an attribute of which is, in particular, Hilbert's space. Once this fact is accepted, the question on whether it is possible to consider the functioning of our brains in the frame of quantum physics becomes irrelevant. Presumably, the problem on the construction of the "theory of everything" (M-theory) requires reinterpretation; it can be now reformulated as the question on the compatibility of postulates of different models of reality. Note that the differences in the physical concepts appear to reside on the interface between the applicability of particular models rather than on the boundary between macro and micro levels. Correspondingly, it is evident that any smooth transition from one type of description to another is impossible since such transition is associated with the change in postulates.

In this volume, we have attempted to consistently justify the assertion that the model of reality that had evolutionarily established at the level of unconscious and then "scientifically shaped" was an object-based one. However, this is not the only way of description. Actually, the problem is in that the characteristics of any other model are translated into object-based concepts with the help of the "object-based" terminology and subsequent reverse reconstruction of the properties using "object-based" notions. This leads to such a mess that the modern "meaningful" part of science has become similar to alchemical treatises describing the structure of matter; correspondingly, the authors felt the need to puzzle out the mentioned problems and hope that this attempt will be interesting to the readers of this book.

The presented material is organized in the following order. First, the fact that if there is no objects in the area of quantum mechanics, then they belong to the corresponding model rather than the reality is proved by case studies of the most discussed and relevant paradoxes of quantum physics. Then, we consider a topological variant in constructing an object-based space that describes the physical properties of an object that are the most verified in science and describable with mathematical relations. Functionality of the proposed construct is tested by deriving the "laws" of conservation of energy and momentum, known from the theory of relativity, in a relativistic form. In conclusion, the criteria are discussed that are to be met to make an object-based model applicable as well as the meaning of some "conservation laws" in physics.

We rely on the known results in the area of psychophysiology concerning our sensory organs and the specific features in processing the flow of sensations to justify that the only possible intuitive model of reality for us is an object-based one. In addition, we analyze the ways to order the primary data into a general outline and to construct a "mental" object-based model of reality at the level of unconscious.

We are aware that this view of "reality" is rather unacceptable in the frame of the current scientific paradigm. While developing the ideas described here, we have discussed these issues with many colleagues who are experts in both physics and psychology. Unfortunately, the lack of understanding was quite common accompanied by the question what for this is necessary. The problems existing in science had long become customary. The experimental part of science is intensively developed; however, the search for new phenomena is kind of random. Perhaps, the new angle of view proposed here will also enhance the development of our knowledge about reality. We strongly hope for this.

NOTES

[1] The authors fully recognize the absurdity and alogism of this assumption since how it is possible to perceive (or measure) something that does not exist.

[2] Italicized by the authors of this book.

CHAPTER 1

The Quantum World and the Problems in Object-Based Interpretation

Суть вечных истин и печали содержит надпись на скрижали.
Но суждено ее понять, когда творец придет опять.

> *The stone tablet tells eternal truths and sorrow. In vain*
> *You try to understand until Creator is back again.*
> *Midnight thoughts*

Abstract: Problems in quantum mechanics are analyzed; this analysis demonstrates that their solution lies beyond the current scientific paradigm, which utilizes an object-based description in a space-time representation. A variant allowing for resolution of the existing contradictions is proposed, which implies a reconsideration of the fundamental principles for reality modeling.

Keywords: Object-based space, Quantum mechanics, Special theory of relativity.

INTRODUCTION

Currently, the comprehension crisis in modern physics is repeatedly mentioned, especially in association with quantum mechanics. Science is developing at such a high rate that it has no time to reconsider its previous axioms; as a result, the new theories, in their essence, are represented by a sum of technologies, frequently poorly matching each other. In this regard, Richard Feynman once noted that no physicist actually understood any quantum mechanics since the reality described by it was completely beyond our common concepts. "Quantum theory is often given as the ultimate argument for the latter vision. Early on, its theorists developed a tradition of gravely teaching willful irrationality to students: 'If you think you understand quantum theory, then you don't.' 'You're not allowed to ask that question.' 'The theory is inscrutable and so, therefore, is the world.' 'Things happen without reason or explanation, according to textbooks and popular

Sergey P. Suprun, Anatoly P. Suprun & Victor F. Petrenko

accounts' [16]. Thus, the decades of futile attempts to understand the new physics using old patterns encourage us to reconsider our most "evident" postulates and turn to the mechanisms underlying these representations. Maybe, they are not as evident as it seems to us? Perhaps, our common sense is no more than prejudice, and it is impossible to construct a new building on the old groundwork? Thus, it is the very life that pushes us to a simple acceptance of the fact that "virtual reality" may well be not the only one possible. While a modern computer can reproduce a plausible simulation, we are completely unaware of the limits of the abilities of our consciousness.

The fact that the solution should be searched for at the level of metaphysics has already become evident. Since this book aims to discuss the way out of the deadlock, it will inevitably leave the boundaries of physics per se in both its informational content and the range of involved issues.

Let us try to clarify why a sort of "instrumental" area of science, such as quantum physics, blows up our traditional concept of reality. We constantly test its "quantum" part for adequacy in experiments but do not cast any doubt on the adequacy of conventional "classical" perception of the surrounding world. A sound argument in the justification of our approach is the assertion that humans as a biological species would hardly survive if our perception of reality were wrong. That is correct, but to what degree is that right? How deep must be the insight into the surrounding world to ensure that the opportunity is not missed? *Homo sapiens* is a product of evolution, which specifies certain requirements in the form of biological programs (instincts) and guarantees their implementation (satisfaction) by the availability of the corresponding tools. However, the next is an intricate and unpredictable process in which the system under constantly changing conditions either proves to be able to exist or not. The organisms living alongside us and having no idea of quantum mechanics belong to a species many times as ancient as we are. They have emerged to be capable of existing successfully with a certain concept of reality, which perhaps is even poorer than ours is. Thus, this argument is most likely not so convincing. Perhaps, the next stage in evolution will allow us to grasp the limits in our understanding of reality.

Let us consider in more detail the psychophysiological basis of an object-based worldview and its layout in a spatiotemporal description.

On Space

In an intuitive manner, our concept of the world around us is based on the perception of reality as a set of objects localized in space and time. Such an understanding is universal, and this concept leads to the known difficulties in interpreting the experimental foundations of quantum physics. However, when contradictions arise between experimental facts and an intuitive perception of reality, which fail to fit the familiar "conceptual" field, it is purposeful to analyze our intuitive perceptual attitudes. Although this is logical and clearly understandable, it appears quite a problem to give up on the evident, namely, with the habitual interpretation of what we see.

To make it clearer, it is necessary to discuss the subject matter of the notions, such as space, time, and object, and trace, at least briefly, their genesis and evolution. The best review for the origin and meaning of the terms space and time has been given by Poincare [17–19]:

"I have shown in "*Science and Hypothesis*" the preponderant role played by the movements of our body in the genesis of the notion of space. For a being completely immovable, there would be neither space nor geometry; in vain would exterior objects be displaced about him; the variations which these displacements would make in his impressions[1] would not be attributed by this being to changes of position, but to simple changes of state; this being would have no means of distinguishing these two sorts of changes, and this distinction, fundamental for us, would have no meaning for him" [17].

Thus, a change in the mental state can be related to either an internal factor or an external one. However, such options are absent for a completely immovable being. Poincare precisely points out that the changes in certain psychophysiological states of a human being can be separately considered and related to his certain movements only if a physical movement of the body is possible. Moreover, movements are understood as a certain succession of *muscle efforts* (which is psychologically experienced as a certain interoceptive process) in combination with tactile, visual, and other efforts. For example, if we move (perform a certain muscular work) around a cube, we, on completing the circle around it and returning to the initial position, have seen four times the same picture, a square. It is experimentally found that the observed figure has certain symmetry and may look differently depending on the observation point (the principles of three-dimensional geometry) yet remain the same entity. This suggests that the "internal" experiences (visual perceptions) are repeated after we have performed a certain work, which allows for interpretation of the "external" medium, independent of our perception. Correspondingly, our motivated and controlled efforts are *movements* in this medium. It is clear that the scheme proposed for the development of spatial concepts is very rough and conditional. In particular, the experience of tactile perception[2] is omitted; however, when rolling a cube in our fingers, we establish the links to our visual sensations, gain the experience of scaling the object by drawing it closer or farther, and form, in general, a consistent database comprising our perceptions of the surrounding world [3].

In his work, Poincare emphasizes the idea that the sole source of our understanding of *external reality* is our sensations ("*internal*" experiences). We **always remain in the space of our sensations;** however, it is considered that they have "different" sources, namely, internal and external. For example, visual objects can be reproduced at our volition ("are accompanied by muscular sensations") by a displacement of the perceiving being in space. Some other percepts cannot be repeated, *i.e.*, they are independent of our muscular efforts. The distinguishing between perceptions and mastering of body control (and, as a consequence, knowledge about the space and its properties) is formed by a person in an intuitive manner from the moment of birth [20]. Exploration of the "ambient" world commences from the distance of a baby's stretched arm. During this period, certain correlations between baby's "internal" sensations start to accumulate; further, these correlations will be established in the metaphor of an object-based perception of the "outward world" in a space-time representation. "External objects, for instance, for which the word object was invented, are real objects and not fleeting with fugitive appearances because they are not only groups of sensations, but groups cemented by a constant bond. It is this bond, and this bond alone, which is the object in itself, and this bond is a relation" [17]. Later, Werner Heisenberg, when describing systems of scientific notions and various fields of science, came to a similar idea: "When one compares this order with older classifications that belong to earlier stages of natural science, one sees that one has now divided the world not into different groups of objects but different groups of connections. In an earlier period of science, one distinguished, for instance, different groups, minerals, plants, animals, and men. These objects were taken according to their group as of different natures, made of different materials, and determined in their behavior by different forces. Now we know that it is always the same matter, the same chemical compounds that may belong to any object, to minerals as well as animals or plants; the forces that act between the different parts of matter are ultimately the same in every kind of object. What can be distinguished is the kind of connection which is primarily important in a certain phenomenon. For instance, when we speak about the action of chemical forces, we mean a kind of connection that is more complicated or, in any case, different from that expressed in Newtonian mechanics. The world thus appears as a complicated tissue of events, in which connections of different kinds alternate or overlap or combine and thereby determine the texture of the whole [21]. Note that the so-called "laws of nature" emerge not without the participation of human, since it is we owing to our natural needs rather than Mother Nature who unites the objects with similar properties and introduces the laws of gas dynamics, hydrodynamics, solid body, and so on.

Thus, the mind assembles a set of repeated correlated sensations, which are interconnected, into an object. There are numerous examples of how the human explores the universe of his/her sensations and how the subconsciousness performs the intricate work of categorization and clarification of the interrelations between different properties of the "outward world" by establishing stable links between "readings of different sensors" of the first signal system. All this represents the mental process that assigns a unit object, event, or experience to a certain class (this class may be represented by verbal and nonverbal signals, symbols, sensory and perceptive templates, social patterns, behavioral stereotypes, and so on). This is first and foremost associated with satisfying the demands specified by Mother Nature, which determines the motivations of behavior. Note that there are certain ways of endogenous stimulation and reward for an individual in the case certain evolutionarily established programs are successfully fulfilled. The potential of *Homo sapiens* is tremendous, and its limits can hardly be imagined. Most likely, it is not the development of these potentials that take place during learning but rather their limitation in the form of selection and fixation of the most likely perception and behavioral patterns in order to provide the fitness to a specific "physical and social environment". Particular object-based space-time paradigms emerge and become implanted at the subconscious level during the so-called sensitive periods in individual development and are further rigidly fixed. As a confirmation, note that teaching an object-based perception to the patients that acquired the ability to see at the age of 16 after the surgery for cataracts may take years. Moreover, some of these patients refuse to learn, complaining of a strongest overstrain of their nervous system.

Turning again to the issue of how the model of space is established, pay attention to the significance of the experiences that are associated with the notion of "displacement–motion". A group of signals dominant in their significance is distinguished in the flux of changing psychic states, while the remaining signals are leveled to the background. According to Poincare, the understanding of a relative position is established, first and foremost, relative to oneself. "But then we may ask ourselves if the relative position of an object with regard to other objects has changed or not, and first whether the relative position of this object with regard to our body has changed. If the impressions this object makes upon us have not changed, we shall be inclined to judge that neither has this relative position changed; if they have changed, we shall judge that this object has changed either in the state or in relative position. It remains to decide which of the two. I have explained in '*Science and Hypothesis*' how we have been led to distinguish the changes of position. Moreover, I shall return to that further. We come to know, therefore, whether the relative position of an object with regard to our body has or has not remained the same" [17].

We omit here the event of how the object "emerged" according to the categorization of the data flux of the first signal system. It is rather problematic to formalize the division of a certain whole initial set into parts, for example, "object–background". In this case, it is necessary to define the concept of boundary and its properties, which has not been so far accomplished in a proper manner. This underlies several paradoxes in mathematics, particularly in the theory of sets [22].

Nonetheless, a human being solves this problem in an intuitive manner by separating a certain class of sensations and using it to form a set of qualities that determine the notion of "object", which is actually identical to the totality of its qualities. Then **the objects are matched to the individual and to one another,** that is, the "external space" is constructed, and the "objects" are placed within this space. In other words, a certain mental map is constructed, which represents the "surrounding world" as a certain projection of the ordered and classified "internal experiences".[4] This mental representation can be further "identified" with reality if the subject himself is excluded from consideration, which makes the overall picture **as objective.** As a result, these experiences are in a way conditionally "subtracted" to construct *an object-based spatial model of* "external reality", which allows the individual to successfully respond to all changes in his/her psychic states in a real-time mode. That is how we create an "objective" worldview.

As for the manipulation of localizing an object in space, Poincare notes, "What does this mean? To localize an object simply means to represent to oneself the movements that would be necessary to reach it. I will explain myself. It is not a question of representing the movements themselves in space, but solely of representing to oneself the muscular sensations which accompany these movements and which do not presuppose the preexistence of the notion of space" [17]. We believe that it is not so simple to describe the mechanism underlying localization of an object as it follows from the above reasoning; however, a conditional character of this process per se is well evident here. In his works, Poincare casts doubts on the physical status of space, its dimensionality, and metrics, which cannot be considered aside from the acting forces. That is why he denied the interpretation of space as a physical object, which is deeply rooted in the general theory of relativity and later led to a number of paradoxes.

However, it should be kept in mind that the notion of space in physics was directly borrowed from mathematics, moreover, in a purely abstract mathematical sense, with its uniformity, isotropy, measure, and so on. Then, it was realized at a certain moment that anything used in a physical model should be experimentally verifiable. The only scientist who presumably understood the meaning of the problem in the right way was Poincare [19], emphasizing that it is impossible to experimentally measure the speed of light propagation over a certain interval (at different points) using one watch. It is only an averaged speed of light passing a certain interval of space in two opposite directions that can be measured; correspondingly, the postulate that the speed of light propagating in forward and reverse directions is no more than a **conventional agreement.** Poincare emphasized that the model of reality should first and foremost meet the demand for convenience and ease of use during communication. Thus, as Poincare wrote [23], we define in a "conventional" manner the rules for the description of our vision of the surrounding world.[5]

On Time

"Проходит все", но неизвестно: в начале мы или в конце,
И сожаленье бесполезно, как эта надпись на кольце.

> *"This too shall pass." Who knows we end or just begin?*
> *Regret is senseless as the words carved on the ring.*
> *Midnight thoughts*

The concept of "time" is discussed in a multitude of papers analyzing its meaning and genesis. Note here that many aspects of time that we experience are indescribable in physical theories, for example, "the flow of time". According to Penrose, "In fact, it is only the phenomenon of consciousness that requires us to think in terms of a "flowing" time at all. According to relativity, one has just a "static" four-dimensional space-time, with no "flowing" about it. The space–time *is just there* and time "flows" no more than space does. It is only consciousness that seems to need time to flow, so we should not be surprised if the relationship between consciousness and time is strange in other ways too" [7].

As early as ancient times, Parmenides, a Greek philosopher, believed that sensory facilities lead to false conceptions, while **the reality is indivisible and timeless.** However, one cannot embrace the unembraceable, as Kozma Prutkov asserted; correspondingly, we have to "split" piece by piece from the entity in order to "digest" it. The succession and orderliness of perception in the background of the memory of receiving system give rise to experiencing the flow of time. Similar to the situation when we create space in order to arrange the items that we distinguish as objects and as a "receptacle" for these objects, time is constructed and serves for us as a "receptacle" of an ordered sequence of our perceptions and experiences. Moreover, this is a discrete sequence since a moment has no content at all, although a parametric description of time in physics is commonly accepted. In this regard, Augustine of Hippo wrote in his *Confessions,* "I said then even now, we measure times as they pass, in order to be able to say, this time is twice so much as that one; or, this is just so much as that; and so of any other parts of time, which be measurable. Wherefore, as I said, we measure times as they pass. And if any should ask me, "How knowest thou?" I might answer, "I know, that we do measure, nor can we measure things that are not; and things past and to come, are not." But time present, how do we measure, seeing it hath no space? It is measured while passing, but when it shall have passed, it is not measured; for there will be nothing to be measured. But whence, by what way, and whither passes it while it is a measuring? Whence, but from the future? Which way, but through the present? Whither, but into the past? From that, therefore, which is not yet, through that, which hath no space, into that, which now is not. Yet what do we measure, if not time in some space? For we do not say, single, and double, and triple, and equal, or any other like the way that we speak of time, except spaces of times. In what space then do we measure time passing? In the future, whence it passeth through? But what is not yet, we measure not. Or in the present, by which it passes? But no space, we do not measure: or in the past, to which it passes? But neither do we measure that, which now is not" [24]. An inadequate description of the patterns and specific features of our perceptions of something that we call reality must sooner or later lead to logical paradoxes.

Once we accept that our perception of reality was initially formed as an object-based one and that this is no more than just a method for categorization of our experiences convenient at a certain stage of evolution, the problems in interpreting quantum mechanics in a "wave-particle" manner become clearly understandable. Find below what is the reason for that and how this can be resolved.

A remarkable specific feature of physics as a science-primarily based on the experiment is the possibility to experimentally verify our concepts. If the "common sense" assumptions contradict facts, then the facts remain. In the last century, quantum mechanics encountered several problems, the understanding of which requires a historical aspect in its establishment. In particular, the discussions of Einstein and Bohr [10] on the completeness and consistency of this theory, on the one hand, have led to its recognition as a tool that efficiently describes reality and, on the other hand, has demonstrated that the essence of this theory cannot be understood in the frame of classical physics. Turn here to work by Einstein, Podolsky, and Rosen (referred to as EPR) published in 1935 [25]. The authors proposed an imaginary experiment that cast doubts on the completeness of quantum mechanics as a theory pretending that it comprehensively and adequately described reality. The main point of this paper was the postulation of two statements quite evident from the standpoint of common sense. First, "If, without in any way disturbing a system, we can predict with certainty (*i.e.*, with probability equal to unity) the value of a physical quantity, then there exists an element of physical reality corresponding to this physical quantity" [25].

Actually, this means that the physical quantity is a property of a certain object that *exists independently of the perception* of an individual (subject), that is, objectively (here, terms "object" and "objective" are not interdependent although they are connected by etymology). Thus, the initial hypothesis is the following: reality is object-based and the *objects exist independently of an observer,* that is, in an objective manner, and their properties exist even *before they are measured.* (The situation when **the properties are primary and the object is secondary** is not considered at all, although current neuropsychology distinctly points to the fact that the "assembly" of any pattern into a single object is the function of our brain [26].)

In addition, EPR meant [25] that cause-and-effect relations between events existed only within a "cone of light" (*i.e.*, the events take place within a time-like interval) specified by the special theory of relativity (STR). Later this statement was formulated as the postulate referred to as the "locality principle". It had to appear because of the need to coordinate an object-based description and space-time representation (this particular principle is kept in mind when speaking about the separability of objects in space-time). The two mentioned postulates distinctly define the principles for modeling reality as an object-based space-time construct. On the other hand, we believe that the object as a given constellation of properties emerges only under certain boundary conditions of its perception; the object itself, in this case, as well as its states, is not specified by its space-time "lock-on". This entails some complications in describing the qualities of an object; an illustrative example here is the electron spin (see later). Therefore, the "classical" representation of an object (permissible classical interpretation for a constellation of properties) is realized by an individual only at the moment of observation. Our emphasis on two statements from this paper [25] (we refer to the first one as the **object-based criterion**) and their significance as the basis for modeling reality within the current scientific paradigm is also confirmed by one of the authors of the work in question [25]. In his letter to Schrödinger on June 19, 1935, Einstein wrote that he would like the separability principle (or locality principle, as is commonly accepted in the scientific literature) to occupy the central place in the paper [25]. The meaning of this statement boils down to that the background for reality is independent events, that is, the processes that allow for manipulations with a certain part of the reality[6] without disturbing its remaining part.

EPR proposed a mental experiment in their work to justify their position; this experiment involves the measurement of the coordinate and momentum of a quantum system comprising two particles. Until that moment, the setup of a decisive experiment in such interpretation was vague. Moreover, its meaning was not even discussed. The fact is that such an interpretation implies verification of the Heisenberg inequality, which deals with the observation process per se, and construction of the model for localization, that is, the arrangement of objects in space-time. Later, David Bohm reformulated the problem by proposing to measure spin, which is a "pure" property [7]. Thus, according to Bohm's proposal, they have to verify the validity of object-based modeling of reality, which was implied by the EPR statement. At approximately the same time, John Bell proved his inequalities, which were applied to the resolution of the alternative, namely, what is correct or, in other words, what is closer to the actual reality—a quantum-mechanical or an object-based reality.

It was **experimentally** demonstrated that neither the postulated object-based criterion nor the locality principle corresponded to the existing reality [27–29].[8] At least, it would be correct when dealing with the quantum phenomena to consider them beyond any space-time categories.

Since we had developed an object-based space-time method for "modeling" reality during our phylogenesis and ontogenesis, we inevitably interpr*et al*l experimental facts within this concept. Moreover, the results of this experiment are predicted using the notion of "probability". The "reduction" of reality into an object-based model allows us to perceive and fix a certain fragment of this reality, regarding its remaining part as likely, probable, accidentally unimplemented, failed, and, correspondingly, not existing in the current reality and absent in past reality[9]. Let us consider the most interesting experiments in quantum physics from this standpoint.

Let us start with the well-known phenomenon, interference, described in almost any quantum mechanics textbook. What is surprising in this case? Usually, the experiment with radiation scattering through two slits is described as follows [30]. We have a source of particles, for example, electrons designated *I* in Fig. (**1.1**).

This source may be a slightly heated cathode creating a single-electron flux. This excludes any interaction between particles so that each particle is alone when it arrives to slit *1* or *2*. A fluorescent screen may be used as a detector; it will record sparkles of light-induced by the arrival of an electron. The spatial distribution of hits along the screen (the x coordinate) is shown as $P(x)$ histogram. Plugging, in turn, slits *1* and *2,* we get distributions $P_1(x)$ and $P_2(x)$, as is evident in Fig. (**1.1b**). These histograms reflect the probability of an electron to hit a certain point with its coordinate x. Note that the sum of these two histograms, designated $P_3(x)$, is completely different from $P(x)$ for the situation with both slits open. This means that the probabilities are not summed[10] in (a) similar to the situation of two exclusive events (for example, heads or tails outcome when tossing a coin). The distribution $P(x)$ for simultaneously open slits is not the sum of distributions $P_1(x)$ and $P_2(x)$.

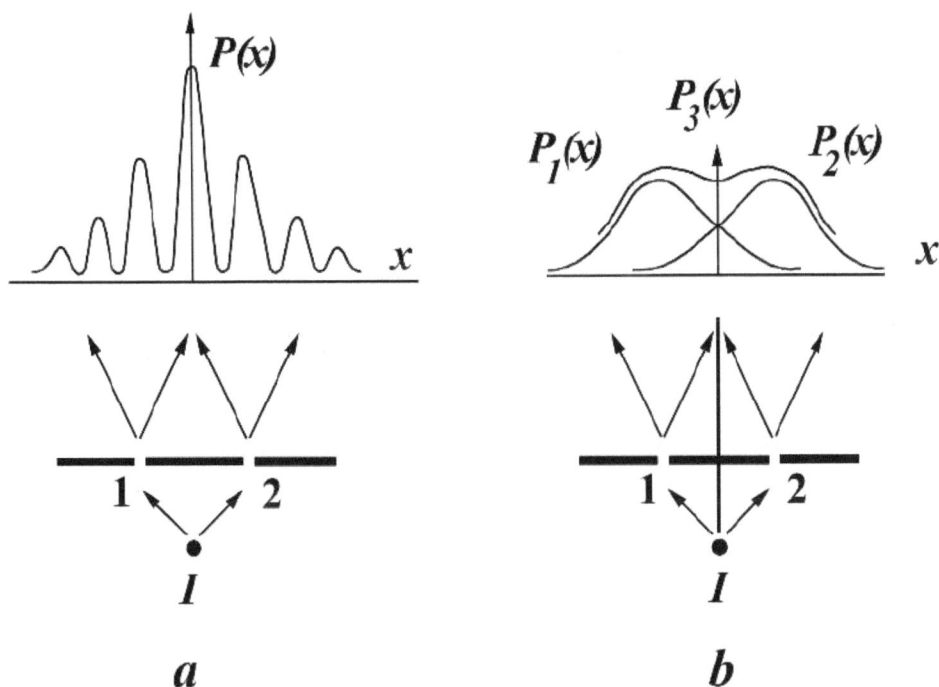

Fig. (1.1). Distribution of the probability for an electron to hit a certain point along the screen in the case of double-slit scattering.

As was noted [30], "By this series of operations, we have shown each electron to have been passing through both slits in the experimental apparatus. But this is utterly impossible. The electron is a particle, and **by the very meaning of this term, we mean something that is at a particular point in space.** A particle can be "here" or "there," but never in two places at once. But the electrons in the experiment of Tonomura *et al.* have apparently managed to do just that." Thus, the authors infer that there is "an experimental demonstration of matter waves", since a correct description of their results requires the concept of the wave function ψ (which is the probability amplitude) to be used.

Nobody has any doubts that reality (exactly reality rather than its model in our consciousness) is **the totality of objects in physical space-time,** namely, the source, barrier, screen, and, of course, electrons. So, to explain how such a "set of objects" can possess such property as "interference", we have to assume that the objects possess wave properties, which are by no means the characteristics typical of objects. Then, the authors do certain mathematical computation, which matches the experimental data. In conclusion, the authors only have to assert, "Quantum mechanics has correctly reproduced the results of the experiment. But has the theory explained to us how indivisible particles manage to pass through two slits at once? It has not. Nothing in the preceding section gives us the slightest insight into how a particle manages to do this. On the one hand, the quantum treatment deals primarily with waves rather than particles. Indeed, a quick glance over the above section will reveal that the very word "particle" plays a little part in the discussion. Aside from a few introductory appearances, the word does not appear until the very close, where the principle was used that $|\psi|^2$ measures the probability of finding "the particle" at a given point. On the other hand, no place made the theory attempt to analyze the particle's path through space. Indeed, quantum mechanics regard the very concept of a trajectory as deeply suspect" [30].

The main problem is not in the fact that we initially assume that the reality is object-based but in that it is actually neither "object-based" nor "wave". This will be clear from further consideration.

The very notion of "object" in physics has long been "under suspicion". In particular, this term has been almost ruled out in the quantum field theory, also known as the standard model: "…we have amply discussed the reasons why quantum mechanics cannot be a realistic theory of the quantum world, which may be summarized in the impossibility to give an objective (*i.e.*, independent of the subject, the observer) meaning to the key notion of wave-particle complementarity. We have also seen that the most problematic aspect of the picture of the quantum world that **quantum mechanics paints us** is the physical nature of the quantum particle, an object that, we should be aware, is quite distinct from the "quantum" of Einstein and Planck. Whereas, in fact, the "quantum" is a particular manifestation of the associated field and does not enjoy any dynamical autonomy, the "quantum particle" according to quantum mechanics, is a well-defined object, much like the Newtonian mass-point, but for the fundamental, and puzzling, difference that the very physical means to define it, by following its trajectory, is in principle unavailable. In this sense, we may well say that the quantum particle is a truly metaphysical object, for no unique objective physical observations exist to give it a real substance. On the other hand, no such difficulties affect the notion of field, which describes in which way a given region of space differs from empty space, where any physical observation yields by definition a null result. *Localization and separability*, two concepts that, we have seen, haunt quantum mechanics, have no fundamental relevance in field theory, for the definition of space and time belongs to the observers through their measuring apparatus (including rigid rods and clocks), and not to the object of field theory, which represents and describes the "physical condition" of the particular region of space-time the observer focuses his attention upon" [31].

The situation with studying the properties of radiation or photons is even more curious. Here we encounter the question of what is a photon in "reality": is it a wave or a particle? This question still has no final answer; however, a number of remarkable experiments have been performed. The point is that the outer photoemission (emission of an electron by a solid caused by radiation) is explainable based on the discrete energy levels in a solid (based on a quantum mechanical model) and purely classical electromagnetic wave without any corpuscular properties postulated by Einstein [10, 30]. When it became clear, they started to search for the conditions that would allow for the unambiguous recording of the photon as a particle[11].

The idea of the experiment proposed for this purpose is as follows: the vapor of calcium atoms is exposed to laser radiation, which is thus excited. When relaxing, the atom passes through an intermediate state with the emission of the first "signaling" photon and then immediately emits the second photon and returns to the initial state. It is necessary to record the signal photon, and only once the signal photon is recorded, all further manipulations are taken into account, namely, the act of passing of the second photon through a semitransparent splitter mirror. After this mirror, the detectors equipped with a coincidence counter are installed on the paths of possible light spreading. The initial hypothesis was formulated in the following way: if photons are particles, the coincidence of counts by different detectors will tend to zero with a decrease in the radiation intensity [32]. Indeed, this experiment recorded the photon as a particle. However, it is necessary to emphasize that this experiment *created unequivocal conditions for observing a quantum system as an individual object.* Moreover, this system lacked the possibility to record interference according to path length difference. Thus, it is no wonder that they observed the properties characteristic of an object.[12]

Making the optic system somewhat more complex allowed Grangier, Roger, and Aspect to create the conditions for observation/non-observation of interference after the above-described manipulations with the photon [32]. Such experiments are referred to as "delayed choice experiments". The fact is that the splitter at the interferometer output (Fig. **1.2**) can be removed or not after radiation has passed the input splitter. Then photon, in one case, would act as a particle and, in the other, as a wave[13]. However, to be always "hot and happening", the photon should always "guess" the next random choice of experimenter and it actually "guessed". This may seem amazing unless we understand that we (or, more precisely, "our" unconscious) by default "forces" the consciousness to make a particular choice for interpreting reality either as an "object" or as a "wave". However, it is actually neither the former nor the latter but merely our own interpretation.

The list of the problems in quantum mechanics is reasonable to continue with the teleportation[14] of quantum state during the decay of the so-called "entangled pairs". In this case, we can speak that the state is "teleported" (including at a velocity exceeding that of light) **to any distance over an indefinitely short time span,** which can be experimentally implemented. Note that this phenomenon is currently applied in quantum cryptography systems [34].[15] The next one to consider is the Aharonov–Bohm effect for particle scattering with a shift in the diffraction pattern in the presence of an electric or a magnetic field in the region **inaccessible** for interaction with the fields. This poses the question on the connectedness of space (and, in addition, on our understanding of the term "interaction") [35, 36].

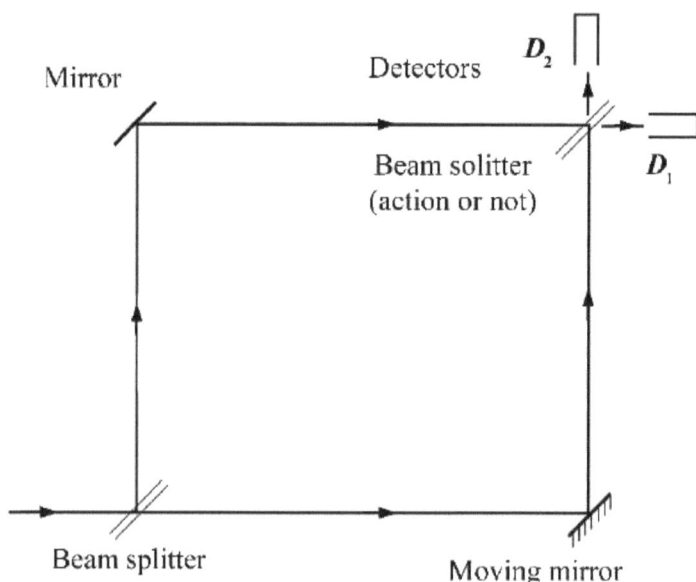

Fig. (1.2). Scheme of an experiment by Grangier, Roger, and Aspect [32].

Another problem in quantum mechanics is associated with a subjective phenomenon of wave function collapse, first stated by John von Neumann: "Indeed, subjective perception **leads us into the intellectual inner life of the individual, which is extra-observational by its very nature** (since it must be taken for granted by any conceivable observation or experiment)" [2]. Analyzing this situation, Penrose concludes that the reduction of the wave function during its measurement is *a real noncomputable physical process,* unlike its *evolution* [37]. Note that Penrose there tries to explain the Consciousness by the existence of such a process. The problem is compounded that the perception of *an individual* on completion of any measurement causes a reduction of the wave function in the consciousness of all individuals! In particular, Eugene Wigner believed that all the unconscious matter evolved in a computable manner. However, once the state of a system appears coupled with the state of a certain conscious being, some physically incomputable process is switched on, thereby leading to the reduction of the quantum system to the perceptible reality, as in the case of the paradox of Wigner's friend [38].

CONCLUSION

In general, the likely inference is that our object-based description method, going hand in hand with a space-time representation, is the particular reason underlying the irresolvable paradoxes when we try to comprehend a "non-object–based" and "non-space–non-time" reality. This is one of the possible ways of modeling reality, yet not the only one. It has its own application area; however, is it, strictly speaking, a scientific approach if our subconscious algorithms underlying perception and construction of a "mental map" for the reality have not been so far comprehensively analyzed? The proof for such a statement is the absence of an adequate ("non-object-based") terminology, which creates certain difficulties in both description and understanding of the corresponding issues.[16]

NOTES

[1] Boldfaced and italicized by the authors of this book.

[2] The role of the human hand as a feedback tool in understanding of reality is tremendous not least because practice is regarded as the only criterion for truth, which frequently misinforms us.

[3] The observed symmetry is described by the theory of groups.

[4] In simple terms, that what we "see" is a virtual picture constructed according to certain algorithms by the unconscious based on our sensations.

[5] For example, an object in one case is regarded as small and in the other, as large

but distant (according to the laws of perspective) although they may well look the same. With approaching the object, its observable size increases but we do not regard them as changing in reality.

[6] Which is supposed to be an object since no alternative is offered.

[7] The properly that permits normalization; see [6].

[8] See [6] for comprehensive description and analysis of experiments.

[9] In this connection, it is rather amazing from an "object-based" standpoint how a quantum computer functions: this implies manipulations with yet unrealized states in order to increase the probability for obtaining a certain result during measurements.

[10] Summing of probabilities is the only variant is the case of an object-based modeling of an outcome for any process since this means that the object-based criterion and locality principle are fulfilled. The very meaning of the probability amplitude is in negation of independence of events.

[11] Note that the Compton effect is also explainable within the concept of wave [30]; see also the references therein.

[12] According to N. Bohr, light is neither particle nor wave, that is, the answer will depend on the setup of experiment [33].

[13] To identify an interference pattern, it is necessary to record a certain number of elementary counts. The specific features of recorded data array are determined by a moving mirror, which changes the optical path difference. See original paper [32] for details of the experiment.

[14] In quantum physics, teleportation is defined as an experimentally proved correlation between measurements in a quantum system that is independent of spacetime intervals between these measurements.

[15] Mention of the quantum state teleportation suggests dwelling on a point

frequently causing incomprehension. Assume a system of coordinates in Einstein's spacetime model. A quantum system of the "EPR pair" type is located there. The first measurement is taken at point (x_1, t_1) and the second, at point (x_2, t_2). In principle, this system can be extended as much as desired and the time span between the successive measurements can be similarly reduced. It is essential here that the first and second events are not random since they are interconnected by correlated measurement results. Correspondingly, it is possible to formally determine the "velocity" of quantum state transition dividing one interval by the other interval. Thus, it is possible to transfer the quantum state; however, this does not allow for transfer of information. Pay attention here to the following specific feature: only a controlled coded signal addressed to an individual can be regarded as information. For example, Alice and Bob may agree that if the answer is positive (Yes), the coin is placed on the table with its "heads" up; otherwise, with its "tails". This is subject to a logical dialogue. However, if the coin is tossed, the result will be probabilistic and uncontrolled. The situation with teleportation is similar when we can at will select, for example, the axis for measuring the spin of the quantum system. However, the measurement result (its direction to one or the other side along the axis) is random and uncontrollable to us (is not determined by our consciousness). This example demonstrates that the STR describes the informational domain in the perception of individual rather than the reality, as commonly believed.

[16] That is why we so frequently use quotation marks, boldface, italics, and footnotes trying to avoid ambiguities or emphasize that the used term fails to completely reflect the meaning of a statement.

CHAPTER 2

Structural Organization of the Brain

"…reality given to us by our sensations."
V.I. Lenin

"We need theory; without theory, we are going to die."
I.V. Stalin

Abstract: The current views of the neuropsychological organization of the brain are briefly reviewed based on the systemic description by A.R. Luria, one of the founders of neuropsychology. The ability of the brain, *i.e.*, to represent reality in a spacetime manner and to construct objects of a totality of elementary sensations, has been analyzed. Extrasensory perception abilities of the brain as well as the potential involvement of brain structures in quantum interactions have been discussed.

Keywords: Consciousness, Functional units of the brain, Neuropsychology, Specific and nonspecific brain systems, Unconscious.

INTRODUCTION

This chapter differs from the others in its content but is absolutely necessary since many further conclusions would look dubious without the outline of the current concept of how our mind "works". Certainly, this is a very brief review of a serious issue. It is difficult to describe the phenomenon, such as our perception of reality, which, as is known, is "given to us by our sensations". The cause underlying the emergence of sensations in the consciousness is commonly related to the impact of various "external forces" on the sensory system. However, the linkage between the bodily processes, being physical in their context, and psychic ones (the linkage that Wolfgang Pauli wished so much to unravel and regarded as the root of all paradoxes in quantum mechanics) is still an elusive puzzle. Presumably, the trouble is that these notions were opposed to each other. In addition, our representation of Reality rather than the Reality itself was separated by the subject-object boundary. However strange it may seem, this contradiction most clearly manifested itself in physics. Thus, the psychophysical problem set as early as antiquity most likely has no solution at all because of the ill-posedness of its formulation within the assumed paradigm. Correspondingly, an overview of the situation in psychology is necessary to discuss this issue.

Sergey P. Suprun, Anatoly P. Suprun & Victor F. Petrenko

The most general systems approach to the neuropsychological analysis of the brain structures was proposed by Aleksandr Luria [26], one of the founders of neuropsychology. He distinguished three major functional units of the brain necessary to implement any mental activity:

(1) The unit for regulation of tone or waking;

(2) The unit for acquisition, processing, and storage of information; and

(3) The unit for programming, regulation, and verification of mental activity.

Unit for Control of Tonus and Waking

Any effective mental activity requires an optimal cortical tone or a state of wakefulness. This is the only condition for an individual to receive and process the information in the best way possible, to remember and retrieve the necessary links, and to plan their activity and control it.

As has been shown by Ivan Pavlov, the excitation processes in the awake cortex obey the law of force, stating that each strong (or biologically significant) stimulus induces a strong response and a weak stimulus, a weak response. Characteristics of the neural processes in this state are concentration, balance of excitation and inhibition, and high mobility of neural processes, making it possible to easily switch between different activity types.

These characteristics of optimal neurodynamics disappear in a subwaking or sleeping state, when the brain cortical tone is decreased. The law of force is violated in inhibitory or "phase" states; as a result, weak stimuli are equalized with the strong stimuli in the intensity of the induced responses (equalization phase), exceeding them and inducing more intensive responses as compared with the responses to strong stimuli (paradoxical phase), or do not induce any response at all (ultraparadoxical phase). In the case of a decreased cortical tone, the balance of excitation and inhibition is disturbed as well as the mobility of the nervous system, which is necessary for normal psychic activity.

The regulatory system that provides the cortical tone resides in the stem and subcortical brain regions and is connected with the cortex *via* positive and negative feedback (Fig. **2.1**). As early as 1949, Magoun and Moruzzi [39] discovered a distinct neural structure in the brain stem; this structure was able to regulate the state of the cortex in a gradual manner (rather than according to all-or-none pattern) by changing its tone and keeping it awake. Since this structure was arranged as a nerve network with embedded bodies of nerve cells connected to each other by short processes, it was named the reticular formation (from Latin *reticulum,* network). A certain part of the fibers of the reticular formation (RF) runs upwards to end in the neocortex. This ascending reticular system plays a key role in the cortex activation and regulation of its activity. Other fibers run down, originating in the neocortex and archicortex to reach the below brain structures. This is the descending reticular system. It puts the subjacent structures under the control of the programs implemented in the cortex and execution, which requires the states of wakefulness to be modified and modulated.

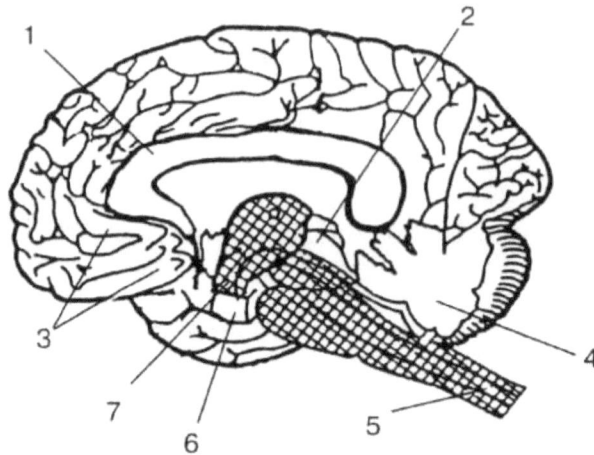

Fig. (2.1). The first functional unit, regulating the total and selective nonspecific brain activation, which comprises the reticular[1] structures of the brain stem, midbrain, and diencephalon,[2] as well as the limbic system[3] and mediobasal[4] frontal and temporal cortex: **(1)** corpus callosum; **(2)** midbrain; **(3)** mediobasal portions of the right frontal and temporal lobes; **(4)** cerebellum; **(5)** reticular formation of the brain stem; **(6)** medial divisions of the right temporal lobe; and **(7)** thalamus.

As has been shown, the first functional unit evokes arousal, increases excitability, sharpens responsiveness, and has a general activation effect on the cortex. The affection of its components drastically decreases the cortical tone, causes sleep, and sometimes even leads to a comatose condition.

The activating RF, the most important portion of the first functional unit of the brain, is referred to as a nonspecific, which cardinally distinguishes it from the overwhelming majority of specific (sensory and motor) systems of the cortex. As is believed, its activating and inhibiting effects uniformly influence all sensory and motor functions of the body, and its function is the regulation of sleep and wakefulness, *i.e.,* the nonspecific background of manifold kinds of activities.

The machinery of the ancient or limbic cortex, occupying the inner (medial) portions of the hemispheres, also belongs to the first functional unit. This includes the thalamic nuclei, the caudal body, and the hippocampus, which emerged to be tightly associated with the system of orienting reflex (response to novelty). The activating and inhibiting functions of the hippocampal and caudal neurons are associated with the most intricate types of orienting reflex, which is of a conditioned character formed during one's lifetime rather than inherited.

Thus, the state of the cortex responsible for active perception and response to external stimuli or the rest state is regulated by a highly organized system with numerous feedbacks. Note that mental individuality, comprising not only the inherited but also the acquired qualities, is preserved at a certain level independently of the state.

Unit for Acquisition, Processing, and Storage of Information

This functional unit resides on the convexital (external) surface of the neocortex, occupies its posterior segments, and comprises the visual (occipital), auditory (temporal), and general sensory (parietal) areas (Fig. **2.2**). The machineries of this unit have high modal specificity and are adapted to receive stimuli that run from the peripheral receptors to the brain, to split the stimuli into a tremendous number of constituent elements, and to combine them into the necessary dynamic functional structures. The components of this functional unit are adapted to receive visual, auditory, vestibular, and general sensory information. This unit also contains the central machineries of the gustatory and olfactory receptions.

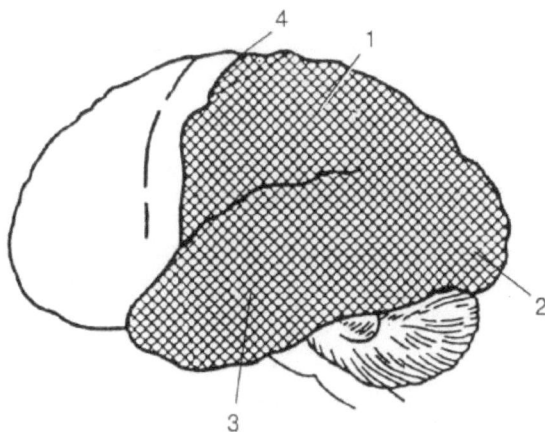

Fig. (2.2). The second functional unit provides the acquisition, processing, and storage of exteroceptive information, comprising the main analyzer systems (visual, skin-kinesthetic, and auditory), the cortical areas of which occupy the posterior regions of brain hemispheres: **(1)** parietal region (general sensory cortex); **(2)** occipital region (visual cortex); **(3)** temporal region (auditory cortex); and **(4)** central gyrus.

In its histological structure, this functional unit is formed of isolated neurons constituting the cortex and arranged in six layers rather than by a continuous nerve network; unlike the machineries of the first unit, these neurons work according to an all-or-none pattern rather than *via* gradual changes by receiving individual impulses and transmitting them to other groups of neurons.

The core of this unit is formed by the primary or projection cortical areas, mainly composed of the neurons of afferent[5] layer 4; a considerable part of these neurons is highly specific. In particular, the neurons of the visual cortex machinery respond only to the narrowest properties of visual stimuli (some neurons, to color tints; others, to the character of lines or direction of motion; *etc.*).

Such hierarchical structure is similarly characteristic of all cortical regions constituting the second functional unit of the brain.

Fig. (**2.3**) shows the primary and secondary areas of the visual and auditory regions and the remaining cortical regions.

The primary or projection areas of the second brain unit are surrounded by the machineries of the secondary (or gnostic) cortical areas built over them; there, afferent layer 4 yields its leading position to cell layers 2 and 3, which lack a pronounced modal specificity. These layers contain considerably more associative neurons with short axons, which allow the input stimuli to be combined into particular functional patterns, thereby implementing the function of synthesis.

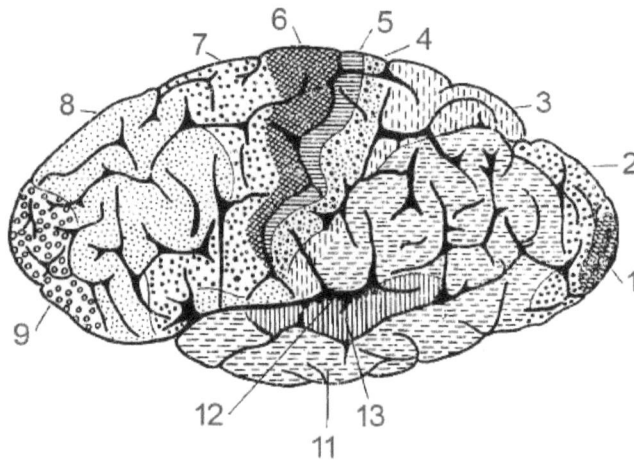

(Fig. 2.3) contd.....

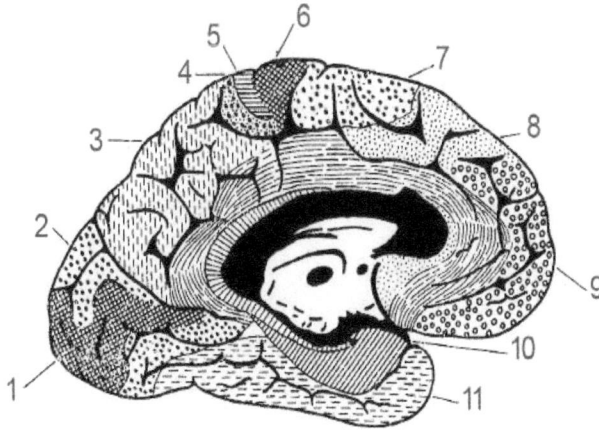

Fig. (2.3). Sensory receptor regions in the cortex of the brain hemispheres: **(1)** visual sensory (primary area); **(2)** visual mental (secondary area); **(3)** parietal; **(4)** intermediate postcentral; **(5)** postcentral; **(6)** precentral; **(7)** intermediate precentral; **(8)** frontal; **(9)** prefrontal; **(10)** olfactory; **(11)** temporal; **(12)** auditory sensory (primary area); and **(13)** auditory mental (secondary area).

Thus, the secondary visual areas (Brodmann's areas 18 and 19; Fig. **2.4**) are built up over the primary visual areas (Brodmann's area 17) in the visual (occipital) cortex. These secondary visual areas convert a somatotopicprojection[6] of individual retinal areas into its functional organization, thereby preserving their modal (visual) specificity, but act as a tool that arranges the visual excitations incoming to the primary visual fields.

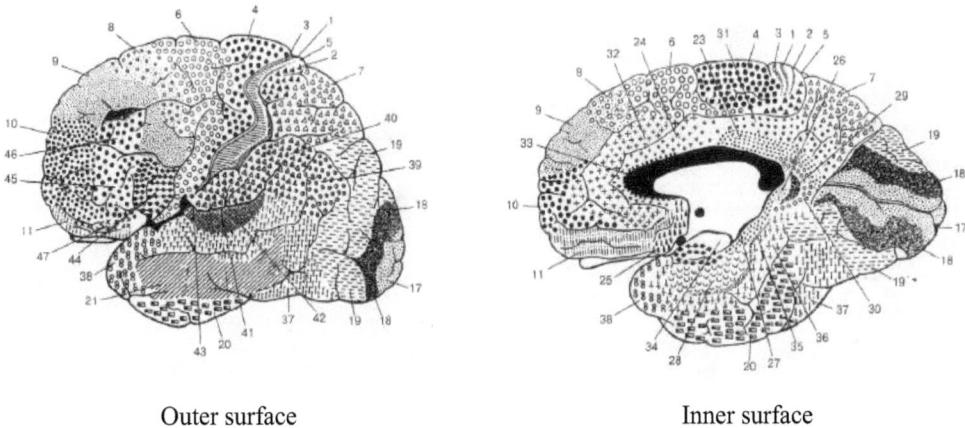

Outer surface   Inner surface

Fig. (2.4). Brodmann's cytoarchitectonic areas and the representation of functions in the cortex of brain hemispheres: **(1, 2, 3, 5, 7, and, in part, 43)** representation of skin and proprioceptive sensitivities; **(4)** motor area; **(6, 8, 9, and 10)** premotor and additional motor areas; **(11)** representation of olfactory reception; **(17, 18, and 19)** representation of visual reception; **(20, 21, 22, 37, 41, 42, and 44)** representation of auditory reception; **(37 and 42)** auditory speech center; **(41)** projection of Corti's organ; and **(44)** motor speech center.

The auditory (temporal) cortex has the same design. Its primary (projection) areas are hidden deep in the temporal cortex in the transverse (Heschl) gyri and are represented by Brodmann's area 41 (Fig. **2.4**), the neurons of which have a high modal specificity responding to only highly differentiated properties of auditory stimuli. Similar to the primary visual field, these primary areas of the auditory cortex have a distinct topography. Several researchers believe that the fibers that conduct excitation from the regions of the Corti organ that respond to high tones reside in the inner (medial) parts *versus* the fibers that respond to deep tones, which reside in the outer (lateral) parts of the Heschl gyrus.

The structures of the secondary auditory cortex are built up over those of the primary auditory cortex. The former occupy the outer (convexital) part of the temporal region (Brodmann's areas 22 and, in part, 21) and are prevalently composed of massive cell layers 2 and 3. Similar to the structure of the visual cortex, they convert somatotopicprojection of auditory impulses into functional organization.

Finally, the same basic functional pattern is also retained in the general sensory parietal cortex. The primary o projection areas (Brodmann's areas 3, 1, and 2) form the framework there as well; they are mainly composed of the neurons of layer 4 with high modal specificity, and the topography has a characteristic distinct somatotopicprojection of individual body segments. As a result, stimulation of the upper regions in this area causes skin sensations in the lower extremities; the middle regions, in the upper extremities of the contralateral side; and stimulation of the points in the lower layer induces the corresponding sensations in the contralateral parts of the face, lips, and tongue.

The secondary general sensory (parietal) areas (Brodmann's areas 5 and, in part, 40) are built up over the primary areas of the parietal cortex as well as the secondary areas of the visual and auditory analyzers. They are mainly composed of the neurons of layers 2 and 3 (associative); correspondingly, their stimulation evokes more complex skin and kinesthetic sensitivity variants.

Thus, the major modal-specific areas of the second functional brain unit have the same hierarchical arrangement, observed in all these areas. Each of them must be regarded as the central cortical structure of a particular modal-specific analyzer. All of them are adapted to act either as a tool for receiving, processing, and storage of the information incoming from the external world or as the brain mechanisms underlying the modal-specific forms of cognitive processes.

However, human cognitive activities never rely on a single isolated modality (vision, hearing, touch, *etc.*) alone. Any object-based perception and, more so, representation is systemic, resulting from a polymodal activity. Correspondingly, it is quite natural that the perception must be based on a concerted work of the whole system of cortical areas in the brain. The tertiary areas of the second functional unit are responsible for the joint work of a set of analyzers; these areas are represented by the overlapping regions of cortical parts of different analyzers residing at the boundary between the occipital, temporal, and postcentral cortices. Their main part is the structures of the posterior parietal cortex, which had developed in humans to almost one-fourth of all structures of the described brain unit. This particular fact suggests that the tertiary areas are specifically human structures.

These tertiary areas of the posterior brain divisions mainly consist of cortical layers 2 and 3 and, consequently, are responsible for the integration of the excitations coming from different analyzers. There are reasons to believe that the overwhelming majority of the neurons forming these areas are multimodal.

The tertiary areas of the posterior cortex are the necessary tools for transformation of the visual perception to abstract thinking and for retention of the organized experience in memory, *i.e.*, they are involved not only in receiving and coding (processing) of information but also in its storage. All of this suggests designating this functional brain unit as the unit for the acquisition, processing, and storage of information.

Unit for Programming, Regulation, and Verification of Mental Activity

The human not only responds to the received signals in a passive manner, but also designs the plans and programs for his/her activities, controls their performance, regulates his/her behavior to meet the plans, compares the effect of the actions to the initial intentions, and corrects the made mistakes. The third functional unit of the brain—the unit for programming, regulation, and verification—organizes the effective, concerned, and focused activities.

The structures of the third brain unit reside in the anterior regions of the hemispheres in front of the precentral gyrus (Fig. **2.5**). The site of exit of this unit is the motor cortex (Brodmann's area 4), the fifth layer of which contains giant pyramidal cells of Betz. The fibers run from these cells to the motor nuclei of the spinal cord, forming parts of the large pyramidal tract. This cortical area is projective and topographically arranged so that the fibers from its upper divisions run to the lower extremities; from the middle divisions, to the upper extremities of the opposite side; and from the lower divisions, to the face, lip, and tongue muscles. The organs of special importance that require the finest regulation are maximally represented in this area.

However, the projection motor cortex cannot function in an isolated manner. All human movements to a certain degree need a known tonic background, which is provided by basal motor ganglia and extrapyramidal fibers.

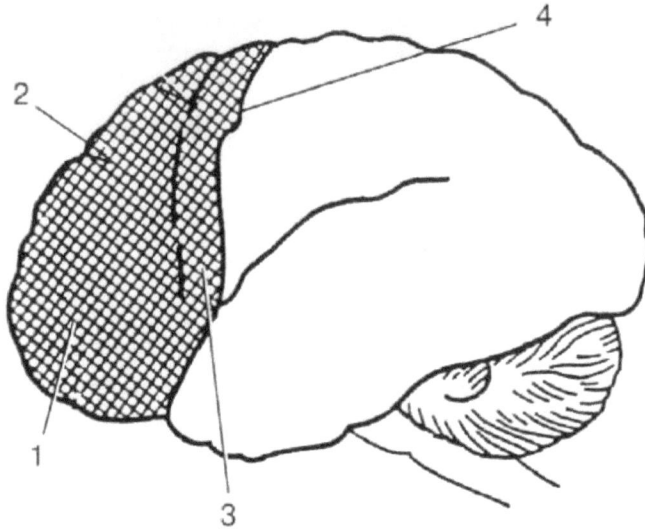

Fig. (2.5). The third functional unit, involved in programming, regulation, and verification of mental activity, which comprises motor, premotor, and prefrontal brain divisions with bilateral links: **(1)** prefrontal region; **(2)** premotor region; **(3)** motor region (precentral gyrus); and **(4)** central gyrus.

The primary (projection) motor cortex is the site of exit for motor impulses. Naturally, the motor portion of the impulses sent to the periphery must be well organized and included into the known programs; only after such preparation, the impulses directed through the anterior central gyrus are able to provide the necessary purposeful movements. The motor impulses are thus prepared by the structures of both the anterior central gyrus and the secondary motor cortical areas built up over this gyrus.

Within the anterior central gyrus itself, the upper cortical layers and extracellular gray matter, formed of the elements of dendrites and glia,[7] are the structures involved in the preparation of motor programs to be conveyed to giant pyramidal cells. The ratio of the mass of extracellular gray matter to that of the anterior central gyrus cells drastically increases with evolution, being doubled as compared with the higher primates and almost fivefold larger as compared with the lower primates.

However, the anterior central gyrus is only a projection area, being an executive tool of the cortex. The secondary and tertiary areas built up over it play the key role and are also arranged in a hierarchical manner with a decreasing specificity similar to the organization of the unit for acquisition, processing, and storage of information. However, its major distinction from the second (afferent) unit is in that the processes here are descending. They originate from the uppermost (secondary and tertiary) areas, where the motor plans and programs are formed and run to the structures of the primary area, which sends the prepared motor impulses to the periphery.

The next distinctive trait characteristic of the third (efferent[8]) brain unit, distinguishing it from the work of the second (afferent) unit, consists in that this unit itself does not contain the set of modal-specific areas representing individual analyzers but rather completely consists of the efferent (motor) type apparatuses and is constantly influenced by the apparatuses of the afferent unit. The premotor divisions of the frontal region are the main players in this unit. Stimulation of these cortical regions induces broad ranges of systemically organized movements (rotation of eyes, head, and overall body or grasping movements of hands) rather than somatotopically limited jerks of individual muscles, thereby suggesting an integrative role of these cortical zones in the organization of movements.

Note also that while stimulation of the anterior central gyrus induces a limited excitation extending only to the adjacent sites, the stimulation of premotor cortical regions affects rather remote regions, including the postcentral areas and, vice versa, the areas of the premotor regions are excited by the stimulation of distant afferent cortical regions. All these facts suggest that the premotor regions can be regarded as the secondary areas of the cortex.

The most essential part of the third brain functional unit is the frontal lobes or, more precisely, prefrontal brain regions, sometimes referred to as the frontal granular cortex. Belonging to the secondary cortical areas, these particular brain regions play a pivotal role in the formation of intentions and programs as well as in the regulation and control of the most intricate human behavioral patterns. These regions entirely consist of small granular cells of the upper cortical layers, with short axons and associative functions. A specific feature of this brain region is its richest network connecting it with both the downstream brain regions (medial nuclei, pulvinar of the thalamus, and other structures) and the corresponding RF regions and the remaining cortical areas. These connections are bidirectional and provide the most advantageous position for the prefrontal cortical areas in both receiving and synthesizing the most intricate system of afferentations running from all brain regions and arranging the efferent impulses, able to regulate all these structures.

The fact that the frontal brain lobes and, in particular, their medial and basal parts possess especially efficient bundles of ascending and descending links to the RF and receive powerful impulses from the first functional unit, thereby "charging" from it by the corresponding energy tone is of fundamental importance. In addition, they may have an especially strong modulatory effect on the RF, shaping its activation impulses in a differential manner to fit the dynamic behavioral patterns, which are directly formed in the frontal cortex.

As has been shown [40], a monkey with excised frontal lobes is capable of simple behavioral actions driven by immediate impressions but is unable to synthesize the signals that integrate perception into a whole situation. In other words, if the signals are not perceived in a single visual field (for example, a rod and a hanging banana), any complex behavioral program that requires a mnestic[9] plan cannot be implemented. Further experiments by several researchers have shown that the removal of the frontal lobes breaks the delayed responses, preventing the animal behavior to follow a certain known internal program (for example, a program based on successive changes in signals). Analysis of these disturbances has revealed that destruction of the frontal lobes interferes with the possibility to inhibit sidetracking stimulations rather than memory. Correspondingly, such an animal is able to adequately respond and perform the tasks at hand only when any sidetracking stimulations are eliminated (complete darkness, sedatives, and so on).

Finally, note a very important function of the frontal lobes in the regulation and control of behavior. A feedback or a reafference mechanism, as an important component of any organized activity, involves the "acceptor of action"[10] [41]; any organized behavior is unfeasible without the acceptor of action. The work of this apparatus is associated with the frontal lobes, which are not only involved in the synthesis of external stimulations, preparation to action, and setup of behavioral program, but also keep track of the effect of performed action and the control of its successful implementation. The destruction of the frontal lobes of an animal deprives it of the ability to assess and correct the made errors, making its behavior disorganized and meaningless.

The intraoperative studies during neurosurgical interventions confirm the above-described systems organization of the brain. As it turns out, stimulation of the primary visual cortical areas induces the simplest visual sensations of the subject, such as colored light, flame, or glowing balls. Stimulating the secondary visual cortical areas makes the subject see more complex objects, for example, butterflies, animals, or human faces. Similarly, a stimulation of the primary auditory cortical areas makes the subject hear individual tones or noises *versus* a stimulation of the secondary auditory areas, inducing fragments of melodies, coherent speech, and so on.

Thus, the primary sensory cortical areas are involved in the analysis of the input modal-specific information and the secondary areas, in its synthesis.

The injury of the frontal lobes interferes with an organized mental activity of the subjects and the concentration of their attention; they lose a critical view of their actions and the ability to realize and correct their own errors; instead of replacing their actions with the necessary variants, they repeat the established stereotypes independently of the set tasks. Thus, the frontal regions play a crucial role in creating the intentions, programs of actions, and their control.

As an example, consider the visual analyzer. In general, the acquired information is analyzed in the periphery by the nerve networks of the eye retina.

First, photoreceptors together form receptive fields, arranged as functional units with the *on* center and *off* periphery (sometimes, vice versa) loaded on the corresponding bipolar cells; in turn, the bipolar cells run to a ganglion cell acting as a kind of comparator for the illumination of the receptive field [42] (Fig. **2.6**).

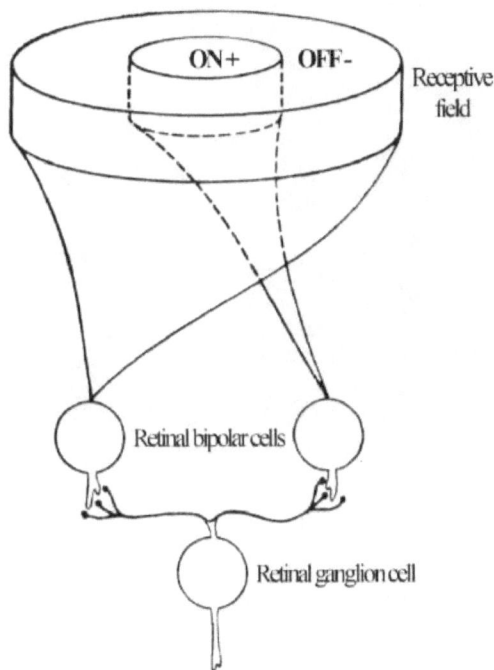

Fig. (2.6). Scheme of the receptive field.

One bipolar cell collects the signals from the *on*-field, connected with the excitation system responding to light, and the other bipolar cell, from the *off*-field, connected with the inhibition system responding to darkening. Thus, the signal reflecting the relative illumination runs to the central nervous system from the ganglion cells, which is illustrated by the well-known illusion (Fig. **2.7**).

Fig. (2.7). Illusion of perception of the central squares of the same hue on different backgrounds.

The color perception is organized in the same manner (the receptors responding to the primary color act as an *on* center and the receptors responding to complementary color, as the peripheral one; together, these colors give gray). The perception of different figures is also similarly organized (Fig. **2.8**).

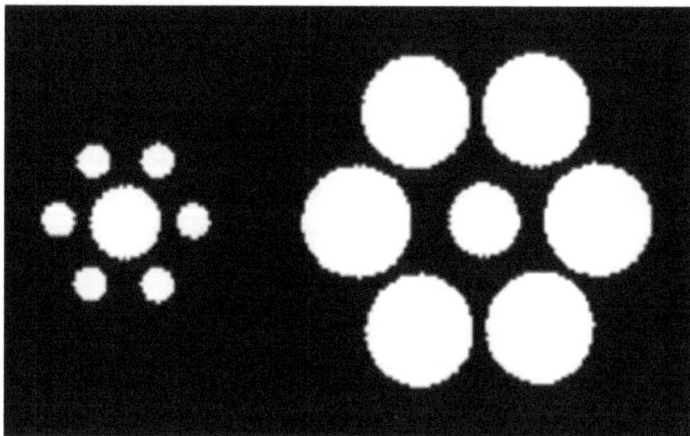

Fig. (2.8). Illusion of perception of the central figures of the same size depending on the periphery.

The nerve networks on the periphery also distinguish straight lines, detect motions, and so on. All these basic characters are transmitted along the optic tract to the primary area of the visual analyzer. Note that the number of neurons in the optic tract is by two orders of magnitude smaller as compared with the number of photoreceptors; thus, the information processing commences in the periphery. Note also that amacrine cells occupy a considerable portion of the retina; these cells have an opposite arrangement of the dendrites and axons as compared with the above-mentioned ones: the excitation is transmitted from the center to the periphery. The fibers of this apparatus that control the eye movements run from the parieto-occipital area of the cortex, responsible for the orientation in space.

The main flow of visual information from the periphery is transmitted to the thalamus. The most important visual centers reside in its posterior part, namely, the lateral geniculate body and the so-called pulvinar. The lateral geniculate body is a well-orchestrated pointwise formatted entrance from the retina to the occipital cortex. The major function of the lateral geniculate body is to prepare the visual information to further processing in the cortex.

As mentioned earlier, neurophysiological data demonstrate that stimulation of the primary visual cortical areas induces elementary visual sensations of the subject, while stimulation of the secondary areas creates representations of integral objects. It is believed that the primary areas fragment and analyze the input modal-specific characters and the secondary areas integrate (synthesize) the information coming to the subject. That is why partial damage of the primary visual cortical areas leads to withdrawal of a limited visual field region. As for the damage to the secondary cortical areas (Brodmann's areas 18 and 19), the subject perceives individual details of an object but cannot unite it into an integral whole. When looking at glasses, such a subject tries to guess what the object is: "What is it? A circle, one more circle… A bar… perhaps, it's a bicycle." It is still rather vague what "sk*etc*hes" of elementary unified image units represented in the primary area are utilized here to assemble an integral, complex, unique image in the secondary area, the more so in the extreme cases shown in Fig. (**2.9**).

Fig. (2.9). An extreme contrast image of a dog (Fig. 3.4 in *Computers: Classical, Quantum and Others,* vol. 1).

It is also amazing that we perceive even flat images as 3D ones (Fig. **2.10**).

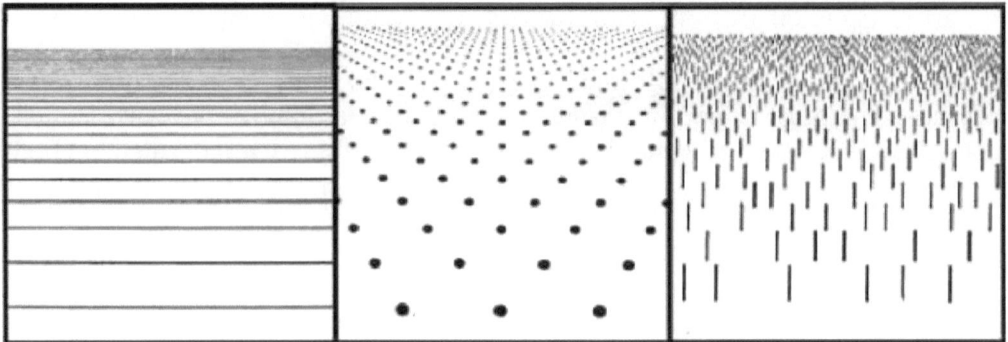

Fig. (2.10). Spatial perception of flat images.

The above briefed currently known experimental results on neurophysiology suggest that the brain is a multilevel hierarchical system with specialized divisions. Simple stimulations of any kind are the mechanism that triggers complex processes of signal conversion into a certain fragment of the picture perceived by an individual as the surrounding reality.

Note that the excitation of certain cortical areas leads not only to experiencing an individual fragment of a certain "reality", but also to the emergence of integrated images. Moreover, this process appears to be completely independent of the first signal system, *i.e.*, of the perception of the surrounding world. Thus, it appears that the human can generate a virtual reality apart from the sensory organs without utilizing the corresponding primary areas.

A number of problems are still to be resolved in terms of the standard scientific paradigm, including the following:

1. We have considered different neuronal systems of the *brain,* which transmit ordinary electric impulses. How do they emerge into modal-specific *sensations of the subject* topologically associated with particular brain regions (a psychophysical problem)?
2. How does the visual scene emerge and is identified in night dreams without any external activation of photoreceptors and sequential operation of the considered apparatuses constituting the second functional brain unit? As is currently shown, sleep and wake are two complementary functional states with reciprocal[11] relationships of their centers, residing in the RF of the midbrain and interbrain. The sleep centers are activated with a decrease in the number of nerve impulses coming to the RF from the peripheral receptors *via* collaterals[12]

and the descending pathways from the cerebral cortex. Excitation of the sleep centers inhibits the waking centers, thereby decreasing the RF activation effect on the cortex.

Dreams are, to a greater degree, characteristic of the so-called rapid eye movement (paradoxical) sleep. This state differs from wakefulness because only some of the brain's activating systems are at play during paradoxical sleep, whereas the most important monoaminergic neurons of the brainstem are inactive. The paradoxical sleep is triggered from the center associated with the pons varolii[13] and medulla oblongata. Commonly, the hallucinations generated by the cortex during this sleep phase are meant when speaking about night dreams.

The brainwaves during the paradoxical sleep have been shown to originate in the stem to propagate upwards through the interbrain to the cortex ("young brain") [43]. The dream narratives, such as a fall from a height, fleeing from a pursuer, holding off an attack, and the attempts to resolve one and the same problem, prove that the structures of the "ancient" brain play an important role in the dreams of the modern human.

Presumably, the information, in this case comes from the unconscious to be converted into an object-based spatiotemporal pattern of a dream by brain mechanisms without the involvement of the consciousness. Anyway, when patients have auditory hallucinations (hear voices), they, for some reason, *are perceived as an external speech* coming from another person, *i.e.*, this speech *is not generated by the consciousness of an individual but rather is perceived by the individual.* In this case, the auditory region of the patient's brain (Broca's area, residing in the lower frontal part of the brain and responsible for speech production, internal speech included) is activated.

Presumably, the primary structures involved in the translation from the unconscious to consciousness are neuronal microtubules of the neuron cytoskeleton, which are able to function at a quantum level [7]. These microtubules are hollow cylindrical tubes with an outer diameter of 25 nm, inner diameter of about 14 nm, and a length of several millimeters. The microtubules ending in the presynaptic axon terminals are associated with clathrins (fullerene-like structures composed of polypeptide three-pointed stars). Clathrins are involved in the release of neurotransmitters and their location allows them to generate both excitatory and inhibitory impulses in the nerve network without the involvement of any receptors.

3. There are several phenomena that are commonly ignored by the scientific community. In these cases, the individual perception, confirmed in an

independent manner, has nothing to do with the canonical brain mechanisms. Find below a couple of examples.

The International Association for Near-Death Studies (IANDS) has been officially registered in the United States since 1981. This Association unites the people who risked studying the problem so unpopular in the academic community and those who have experienced such situations.

In 2001, Pim van Lommel, a cardiologist from the Netherlands, and his team examined near-death experiences of 344 patients resuscitated following a cardiac arrest in 10 Holland hospitals [44]. One patient had a typical out-of-body experience; he reported that he could observe and remember the events during his cardiac arrest. The hospital staff confirmed his statement: this did not look like hallucinatory or illusive experiences since the recollections matched the real and verifiable events rather than imaginary ones [44].

The IANDS achieves have accumulated a lot of evidence of mystical experience. For example, one of the most outstanding cases is known the case of Maria, an immigrant worker who experienced a cardiac arrest. In 1977, she was brought to a Seattle hospital. Maria was successfully resuscitated; then, she informed that her soul left her body during the clinical death and, for a certain time, was floating near the ceiling of the operation room. Then her soul breezed out of the building, lifting higher and higher, and noticed a tennis shoe on a third-floor window ledge, which she described in some detail. Kimberly Clark Sharp, a social worker, went to the window Maria had indicated and not only found the shoe but said that the way it was placed meant there was no way Maria could have seen all the details she described from inside of her hospital room.

4. The papers on the studies into extrasensory perception, which also fail to match the current concept of how the brain works, ever more frequently appear in the scientific literature (see, for example, [45]).

In all these cases, the analogy of the body to a virtual reality suit and of the brain to a processor unit transcoding the input information to an object-based representation seems more pertinent than the standard theory of reflection of external reality.

According to von Neumann [2], the observer in quantum mechanics implements the reduction of the wave function, thereby providing a selection of one possibility among a multitude of others. In a sense, perception creates the "classical reality" of consciousness described by classical physics, not even realizing it. Consequently, this work can be done only by unconscious mechanisms.

CONCLUSION

Note that the subject can be incorporated into the physical theory by matching the physical and psychological paradigms. In terms of psychology, what becomes a fact of the consciousness emerges from that what is not the consciousness, *i.e.*, from the unconscious. In physics, the same process is not related to the unconscious but to the so-called "external world". Perception may be regarded as the process of translation of certain finite content from an open system of the unconscious to the system of individual consciousness. Since the state of a system is defined *via* the content of this system, a change in the content of consciousness results from a certain change in the state of a subject; in turn, the subject relates this to a change in the state of "external medium" (actually, alienating the state of the subject to the state of the physical system). In terms of the systems approach to the psychophysical problem, the external medium, according to Occam's razor principle, is an extra entity. In fact, this is a second signal system (according to Pavlov) semiotic description of the content of our consciousness and its logical formalization. Note that the information we get about the "external" reality is exclusively of a signal (Pavlov's first signal system) character, while the qualia emerge only within the individual consciousness system by means of the representation tools formed by Evolution. Moreover, this representation could be not only of an object-based and spacetime type, common for us, but also, for example, of a spectral type in an infinite-dimensional Hilbert space, which better matches the formalism of quantum mechanics, able to most adequately describe reality, and, consequently, better fits the "language" of unconscious. The impossibility of such direct reality representation (the "gate" to unconscious) is associated with stable innate and acquired object-based forms of reality representation in the consciousness, actually defining the boundaries of *consciousness*. A limited character of the object-based (local) spatiotemporal representation and its inability to fill in the whole picture (content) of reality is demonstrated by numerous quantum mechanics paradoxes, such as the uncertainty principle, quantum teleportation (instant teleportation of a quantum state to any spot of the Universe, which Einstein yet failed to accept), and quantum eraser.

Note that consciousness has its individual observer's reference frame (physical, biological, and social), which in no way refers to the unconscious since the choice of unconscious immediately transforms possibility into "reality" *in any* individual consciousness (recall the paradox of Wigner's friend, a kind of sequel of the paradox of Schrödinger's cat). In this regard, the studies by Carl Jung (written with the liveliest participation of Pauli, a well-known theoretical physicist and a Nobel Prize winner) on the collective unconscious and synchronicity, able to explain many of these paradoxes, are of particular interest.

Actually, Jung and Pauli attempted to decompose the psychic reality, which evidently did not fit the object-based paradigm, at least because of the specific formalism at the level of mathematical apparatus, in a systemic manner, although they still tried to interpret the results in an object-based paradigm. Pauli hoped to link the quantum reality and consciousness *via* the Jungian collective unconscious and thereby resolve the artificial psychophysical problem and paradoxes of quantum physics. In the psychology paradigm, the unconscious iswhat physicists refer to as "external reality", which is the source of our sensations, the limits of which we cannot leave. It is just senseless to separate and then link the consciousness and rationally constructed tangible "external reality" *via* the interaction in terms of resolution of the psychophysiological problem.[14]

However, it is reasonable to examine the effect of individual perception as the observer's reference frame on the state of the quantum system since, according to Poincare [17], it reflects the state of the subject. Here, a psychosemantic approach may be very useful, due to a certain similarity between the mathematical formalism utilized in both psychosemantics and quantum mechanics.

NOTES

[1] The totality of different neurons that reside along the brain stem and activate or inhibit different structures of the central nervous system, thereby controlling their reflex activity.

[2] The major function of the diencephalon is not just provision of the communication with the outer world but the regulation of cortical tone, drives, and affects. The overhelmingmajority of the limbic structures are involved in the functional organization of emotions, which suggests their effect on the corresponding vegetative changes controlled by the hypothalamus.

[3] A complex of structures of the endbrain, interbrain, and midbrain, which is the substrate for evoking the most common states of the body (sleep, wakefulness, emotions, motivations, and so on).

[4] Structures of archicortex, paleocortex, and periallocortex, which retain tight connections to the nonspecific thalamic nuclei, basal ganglia, and other nonspecific structures, for example, the reticular formation and diencephalic nuclei. In Luria's concept on the three functional units of the brain, the mediobasal divisions of the cerebral hemispheres belong to the first (power) block, the main function of which consists in the regulation of the level of activation (tone) of the brain.

[5] Afferentation (from Latin *afferens,* bringing along or with) is a constant flow of neural impulses conveyed to the central nervous system from the sensory organs, which receive information from both the external stimuli (exteroception) and internal organs (interoception).

[6] A somatotopic representation is understood as the phenomenon of self-projection of human organs and systems onto local sites of the own integument or internal organs with a more or less strict preservation of topographic proportions.

[7] Glia is the structure of the nervous system formed by specialized cells of various shapes that fill the spaces between neurons and capillaries, accounting for up to 10% of the brain volume.

[8] Efferent (from Latin *efferens,* carrying outward) nerve fibers conduct the signals from the central divisions to the below and peripheral nervous system divisions (also, efferent signals). The synonym is centrifugal (from Latin *centrum,* center and *fuga,* flight).

[9] Mnestic activity is a specifically human feature, absent in animals, namely, the ability of brain to fix and memorize any information to reproduce it at the right moment. When an animal develops a skill or a conditioned reflex, a certain type of activity is induced, which is preserved by repetition.

[10] The acceptor of the results of action forms the neural mechanisms making it possible to predict the characteristics of the result necessary at this very moment and to compare them to the parameters of a real result received by the acceptor *via* reafference. This particular apparatus gives the unique possibility for the organism to correct the behavioral errors and make the behavior maximally effective.

[11] Reciprocity (from Latin *reciprocus,* meaning returning, reversed, or mutual) is the physiological term describing the interaction of the structures of the nervous system with mutually concerted but oppositely directed coordination of their function; in this process, the activity of one structure (neuron, nerve center, or nerve network) as a rule decreases or inhibits the activity of the other nerve structure and vice versa, thereby leading to the corresponding change in the function of the organs and tissues regulated by these structures of the nervous system.

[12] Collaterals are sideways or bypasses of blood circulation, blood vessel branches, which provide the blood supply in addition to the corresponding main vessel in the case of its thrombosis, embolism, and compression, ligation, or obliteration of vessels.

[13] Pons cerebri or pons varolii (Latin) is located above the medulla oblongata and is involved in sensory, conducting, motor, and integrative reflex functions.

[14] The psychophysiological problem (part of psychophysical problem) is the question on correlation (interaction) between the human body and psyche.

CHAPTER 3

The Principles of Semantic Modeling

"Das Bild ist ein Modell der Wirklichkeit"
"A picture is a model of reality"
Ludwig Wittgenstein, Tractatus Logico-Philosophicus

Abstract: Psychosemantics studies the genesis, structure, and functioning of an individual or collective consciousness and its major element, the meaning. The last can be fixed in the words of a natural language as well as signs, symbols, pictures, and so on in their meaning invariant to different individuals of the same culture in a socially scaled sense. To have a chance to compare sensations and to simulate (represent) the surrounding reality, an individual needs to "construct" a certain semantic space (mental map of reality) and define metric and system of coordinates. Definition of semantic spaces is the mathematical method to construct a mental map as a system that simulates an "objective reality" in terms of mathematical structures (arbitrary sets with the relations defined on them).

Keywords: Mental map, Object-based modeling, Psychosemantics, Rigidity and intensity of sensations, Semantic space.

INTRODUCTION

Since the only cognitive method for representation of the "external reality" for us is a mental one, we have to very attentively consider the mental simulation of this reality. In our view, the only way is to analyze the process of perception and the algorithms for the construction of a sign-based description of reality by man. This process provides the possibility to revise a wide range of scientific problems by using semiotic and psychosemantic methods in the analysis of the methods for the construction of object-based reality. The most efficient analytical tool that makes it possible to distinguish the consistent patterns in the collection, storage, and transformation of the totality of characteristics that we define as an object is the mathematical modeling of the "mental map" of consciousness, *i.e.*, the mental image or mental model of reality. Evidently, the subject at this stage plays significant role in defining what we refer to as "objective" reality and must somehow be incorporated into the general worldview, including the "physical" landscape.

Sergey P. Suprun, Anatoly P. Suprun & Victor F. Petrenko

The construction of *semantic spaces* is the method for the mathematical formulation of the mental map as a system that simulates "objective reality" in terms of mathematical structures (arbitrary sets with the relations defined on these sets). As is shown below, the formalization of semantic analysis has allowed, as a consequence, for the discovery of several relations coinciding with the basic physical laws but extending far beyond the purely physical area; this indirectly confirms the adequacy of the approach to the general description of reality taking into account the new role of the subject in scientific knowledge. This made it possible to formulate the scope of semantic analysis: How the method used to reflect the world by the subject influences our knowledge about it. Semantic analysis implies the existence of the common nonspecific regular patterns within any science that are associated with "topological" characteristics of the mental map and arise owing to the specific features of reality representation in this map.

Psychosemantics studies the genesis, structure, and function of an individual or collective consciousness and its major element, the meaning. The meaning can be fixed in several forms, first and foremost, in the words of a natural language as well as signs, symbols, pictures, body language, kinds of ritual behavior, and so on in their meaning invariant to different individuals of the same culture. "…The awareness is associated with the existence of certain permanent sign forms that carry the meanings allowing the world to be represented to a subject" [46]. Charles Sanders Pierce gave the most general definition of designation as "the translation of a sign into another system of sign". Roman Jakobson, a well-known linguist, averred in this regard: "How many fruitless discussions about mentalism and antimentalism would be avoided, if one approached the notion of *meaning* in terms of translation, which no mentalists and no behaviorists could reject" [47]. Let us demonstrate that according to Pavlov, the translation from the "first signal" system of signs, intended for "domestic use", to the "second signal" system, intended for interindividual communication, actually gives birth to the *meaning* of a sign.

Recognizing a "signaling" nature of our sensations in their origin intended for orientation of individual in the reality residing "beyond" sensations, we should also recognize a *model* and *sign-based* character of our view of reality. First and foremost, note that *this is a one-sided* "communication" despite "reafferentation" [48], namely, "Nature" $\rightarrow$ "individual Consciousness". Moreover, we receive not the "ready-for-use" signs but rather a "raw" process $S(t)$, which should be decomposed into elementary components $s_i(t)$ to extract the elements that carry the information necessary for us (to recognize them as *signals*). Thus, the particular "signals" *significant* for an individual[2] are selected and used for further fine-tuning only as a result of a series of autonomous manipulations after *the external* "exteroceptive," and *internal* "interoceptive"[3] processes *are correlated,* and the "correlates" are selected. The next stage is filtering off the "noise", *i.e.*, the currently irrelevant processes, followed by memorizing, matching with other signals, and so on. The *first-level signs* or *sensations* are constructed of the selected signals only after all these stages are completed.

We are forced to use the established terminology; however, considering this problem more strictly, we may not refer to the processes of "external stimulation" as signals since Mother Nature *had no intention at all to tell us anything.* We "ourselves" choose[4] the components significant for satisfying our demands and then *interpret them as the signals* that are *meaningful* for us in a particular situation.

Evidently, in order to represent *qualities* and their *intensities,* we need to recognize the type of functions $s_i(t)$ as well as the domains of their definition and variation. Correspondingly, an instantaneous value, $S(t)$, has zero informational content.[5] For example, it is minimally necessary to separate a tone when recognizing a sound, *i.e.*, to perform spectral processing of the signal over at least two periods.[6] That is why any sensation only looks like an instantaneous event but actually comprises the process of a time-wise "convolution of a signal" *with a certain duration* Δt. The impression that a sensation is instantaneous[7] appears because the transition of an unrecognizable *signal* to a recognized *sign* is a *qualitative transition* (an instantaneous event when *the sign appears in consciousness*[8]). An integral content stands behind a sign, and this content cannot change in a gradual manner (it is impossible to be "a little bit pregnant"). The change in form in the course of a process can be continuous, but the change in the content of the process is always a "jump".

In order to have a chance to compare sensations and simulate (represent) the surrounding reality, an individual needs to "construct" a certain mental space[9] and define its *metric* and *system of coordinates*.[10] The latter means the definition of "neutral" or indifferent levels of sensations ("zero points") and the setting of the *primary oppositions* of a "favorable–adverse" type, which are used as references to assess the intensity and sign of sensations. Actually, it is also possible to consider the evolution of mental spaces in connection with the evolution of the nervous system and receptive machinery, which allow for various types of comparisons, from the simplest to the most complex; however, this, in essence, has been already done by mathematicians when studying the spaces of different types [49].

In the next stage, it is possible to perform *object-based modeling* of reality, *i.e.*, to construct a *mental map,* by separating the points (or areas) in the mental space as stable "sets of sensations" that are significant for the survival of an individual.

Actually, the *mental map* is a limited *semiotic model* of reality beyond sensations [46], which allows an individual to orient in this reality and plan (program) beforehand the actions in spacetime directed towards the satisfaction of the relevant needs. The adequacy of this model is verified according to the efficiency in satisfaction of these needs, *i.e.*, according to the "survivability" and well-being of an individual.

It is evident in terms of mathematicians that *the objects in their construction* already have a *certain meaning* for an individual, being automatically related to his/her demands (since the "external" signals[11] were *distinguished as "correlates" of interoceptive signals,* reflecting the needs of the organism) and actually play the role of the *tools for satisfying these demands* (direct or indirect). Matching the objects *to each other* on a mental map now makes it possible to construct the secondary more *abstract relations* and oppositions that determine the *signs of the next level,* describing the object-based *situation, operations, and processes.* In any case, however, all these signs in their origin *are not intended for bilateral communication,* and the mental map of consciousness *is unavailable to perception by other individuals,* being indefinable in spacetime as a "real" object. In a strict definition of sign, the sign is an *object-based tool for communication,* while sensation is a mere *subjective marker* for a component of "stimulation" significant for an individual.

An interindividual communication ("individual" ↔ "individual") requires valid signs relying on the so-called "external *object-based* means" definitely represented in mental maps of other individuals, *i.e.*, available for perception by *others*. Actually, the speech emerges as a *naming* of the significant points that determine an object on a mental map via relating to certain "external" processes implemented by the *individual himself/herself in his/her environment* (speaking, writing, and so on). Other individuals interpret them as *signals,* and they are again transformed into "sets of sensations", *signs of the second kind,* since they *are not "necessarily" natural* but rather *determined by a "social contract".*[12] They are initially "anchored" to **signal** stimuli and are a **communication** tool. Note that the space–coordinate nature and meaning component of a *sign–name* is concealed.[13] In the structure, what stands behind the sign is always limited[14] and attached to a certain system of coordinates and the relevant needs, while the name creates a sort of illusion of absolute self-identity of an object (the independence of its meaning from the observer's reference frame and demands). For example, a change in the mental state (transition to anxiety) inevitably changes the "indifferent" point specifying the origin of coordinates and makes an individual emit signals of another type or intensity. Undoubtedly, this adaptive change in the frame of reference influences the *absolute **meaning** of an object.*

Thus, by experimentally modeling a "named" object as a second-level sign in the *space of qualities,* we will inevitably obtain its *different coordinate representations* for different individuals in different mental states. Actually, we will obtain different lexical meanings of a word–name in the language or different *states* of an object, which characterize the perception of the object by an individual in different mental states or situations.[15]

Since an ethnic group creating its language usually lives under more or less standard conditions and biologically belongs to the same species, both interindividual communality and stability in the meanings of the words in this language ensue from an objective generic similarity of their mental maps. Naturally, the similar *states most probable for the population* are utilized for the establishment of the common (lexical) meanings of a word; otherwise, the very chance of interindividual communication is lost.

Another difficulty in psychosemantics is the definition of the meaning of a word. Since an object is described by a set of qualities *differing in their value* for satisfying *different needs,* it is evident that the *meaning of an object (for which particular purpose* this object is necessary to an individual) will also change with updating and satisfying these needs.[16] However, the meaning is a more individualized and flexible characteristic of the word. The meaning is reflected in the *rating of preferences of its "constituent" properties* under different conditions rather than in its absolute definition in the space of properties (*i.e., in its meaning*). It is evident that a change in the value of qualities also changes the preferences of different objects in the body of a certain class,[17] which can form the background for a formal construction of motivating vectors in the semantic space.

Thus, the psychosemantic modeling of the mental map of an individual implies the following stages:

1. Construction of a semantic space[18]*adequate to* the space of the mental map of an individual (or of a particular mentality) that allows for *description of the objects of any type;*[19]
2. Definition of the object as a superposition of its possible meanings and the probabilities of their actualization (*i.e.,* taking into account all most probable interpretations when being perceived by a particular mentality); and
3. Construction of the motivation space based on the semantic space and definition of individual meanings for stable denotations of objects.

Find below the explanation of the above statements in more detail. Any organism is a complex system that perceives the outward world. Most likely, the more complex the system, the larger the toolkit it can use to "tune" itself to reality. The price for this "openness" is a higher vulnerability; however, this brings about new opportunities for development. The ability of an organism to survive and adapt is the response to the environment, which, however, does not imply any survey of reality on a scale beyond the challenges set to this organism by Mother Nature. Presumably, the only species, *Homo sapiens,* is fortunate to cross "the innermost line of defense" against the surrounding world, in particular, utilizing the toolkit that considerably increased the scale and number of the channels of perception. However, the psychophysiology of humans, or, simply speaking, the machinery for data processing, has not changed since the time of troglodytes. Yet, this is only one of the arising problems.

A biological system needs to order its sensations, arranging them into a certain integral linked picture, that is, to integrate the incoming signals into a "representation" of the environment. This is what we refer to as the model of reality, and this is how the ability of an adequate response and purposeful planning of our own behavior emerges. Still, the questions are whether this pattern is cognitive and what it is like.

Presumably, the man realizes, although in part, that he lives in a world of his own "illusions". This issue is frequently discussed, at least in the area of psychology, although in a rather abstract way. However, this representation of reality became paramount when our biological species had reached a social level of development. The emergence of a need in communication brings about the problem of how a representation is translated into a sign. This problem boils down to partitioning an integrated representation constructed by the unconscious into individual elements, distinguishing the significant, and transmitting the content as certain resulting information. This is how we proceed to the semantic space filled with objects and understandable in a certain social medium. This conceptual space is formed when an individual is fostered, trained, and educated during their entire life.

Once we have accepted the fact that the semantic space exists, it is tempting to construct a mathematical model and, first and foremost, to devise a method allowing an object to be represented as a body of its characters. At first glance, this is a complex problem, demanding a stepwise definition of the methods to solve it. An object is completely characterized by the set of its qualities, which are naturally the elements of mental space. However, these elements are derivatives of our cognition verified by practice and experiment. According to Dmitri Blokhintsev [53], "Our knowledge of the external world relies on this particular feedback and its forms can be most diverse. It is clear that is the feedback fails to correctly work, *i.e.*, fails "to reflect reality", neither an individual organism nor the whole community having this defect can exist for a certain long period. However, the methods of the reality "is reflected" can be rather diverse too. In particular, night vision devices convert infrared light, unperceived by the human eye, into visible light; the color of an object can be transmitted by both a light signal (a photochemical effect of light on the eye) and a sound signal if the light is converted into sound using photocells and the sound is passed to auditory organs."

The question is whether the description of an object as a body of certain elements is a unique representation. Let us start by considering a conventional problem in factor analysis related to distinguishing the principal factors in a statistical data array that characterize a certain phenomenon. Evidently, the set of the qualities to be considered may well result from the fact that particular characteristics are easy to observe and record because of, say, the availability of the corresponding tools. For example, we are interested in the class of objects, such as "tables", and we have the tool, such as a "ruler". Thus, we can measure some characteristics of our object, for example, height, length, and width. In a general case, we get a set of data for each table in the form of a matrix, with each row representing the corresponding measurements of an individual object. For convenience, each "quality" in the further analysis is by some means normalized, for example, to the maximum value $z = x/x_{max}$ or as

$$z = (x - \bar{x})/\sigma, \quad \bar{x} = \frac{1}{n}\sum_1^n x_i, \quad \sigma = \sqrt{\sum_1^n (x_i - \bar{x})^2 / (n-1)}$$

where x_i is the measured value; x_{max}, maximum value; $\bar{x}$, mean; and σ, root mean square deviation. This allows us to deal with dimensionless values. The correlation coefficient for the set of variables x and y is put down as

$$r = \frac{\sum_i (x_i - \bar{x}) \times (y_i - \bar{y})}{\sqrt{\sum_i (x_i - \bar{x})^2 \sum_i (y_i - \bar{y})^2}}$$

Marking our objects (tables) as dots in a three-dimensional space (height, length, and width), we get a Gaussian "cloud" resembling an ellipsoid. The correlation matrix computed for this case will demonstrate that the characteristics we selected are not independent. The further mathematical operations are reduced to the search for the principal directions along which the vectors of characteristics from bundles. Since the direction of our interest must pass through the bundle's center of density, the vectors forming the bundle give the largest projections onto this

particular direction. The bundle density is determined by the total sum of the squared projections[20] onto a particular direction.

However, it is possible in our example to select two orthogonal directions that explain approximately 90% of the variance of experimental observations. This is associated with the fact that the tables are tailored "for humans", that is, the table must provide convenient work when a person is sitting. Then tentatively speaking, the key factors are "body height" and "arm length". Now, please take into account that our survey could include coffee tables, a table in a conference room, or a Japanese dinner table (and they used to eat sitting on the floor in Japan). Thus, we need to introduce additional factors to explain the remaining 10%; however, these two mentioned factors would be, in many cases, sufficient to define the essence of our object.

Thus, the characteristics that we initially selected for a certain part of reality distinguished as an object are helpful for "supervision" but are "nondescript" since they fail to reflect the purpose of this object. The performed analysis makes it possible to "compress" the information by distinguishing the principal components that reflect the content of the examined class of objects with a concurrent reduction in the number of characteristics.

This example is interesting in that we deliberately search for the key factors that define the object in this case. Conceivably, our unconscious acts in a similar way. Now let us try to distinguish the main patterns in its functioning. First, we intuitively normalize the characteristics of objects "to the ideal"; correspondingly, this can be mathematically represented as a unit orthonormal basis. This basis will contain both the positive value of quality and its opposite via the angle of the radius vector. Then, the intensity (degree) of characteristic is determined by the angle. For example, selecting the hue of a color by mixing or diluting dyes, the intensity can be changed according to the following equation:

$$\varphi_3 = arc\sin\left\{\frac{(R_1 + R_2)\sin\varphi_1\sin\varphi_2}{R_1\sin\varphi_2 + R_2\sin\varphi_1}\right\} \; and \; R_3 = (R_1 + R_2)$$

where R is rigidity or, in a special case, just mass. Thus, it is evident that if $\sin\varphi_1 = \sin\varphi_2$, *i.e.*, the intensities of the qualities of objects are equal before merging, then $\varphi_3 = \varphi_1 = \varphi_2$. However, some characteristics of the object will be summed up, for example, mass. Thus, an adequate semiotic representation of a characteristic in a language requires that not only its *quality* and *intensity* are given (φ_i angle is specified), but also its *rigidity*,[21] the length of the vector that describes the object, *i.e.*, **the resistance of the characteristic to a change in its intensity.** This particular representation of objects can be referred to as a *semantic* representation.

Presumably, the additive characteristics must "be filed" to the space of properties, which corresponds to the operation of direct multiplication in topology. Consider an example in the area of spatial perception, which is directly associated with the experience of "reachability" of objects, and the illusions frequently accompanying this situation. According to Schrödinger, the specificity in the formation of our spatial experiences via certain physical efforts lies in the following: "Hundreds and hundreds of manipulations and performances of everyday life had all to be learned once, and that with great attentiveness and painstaking care. Take, for example, a small child's first attempt at walking. They are eminently in the focus of consciousness; the first successes are hailed by the performer with shouts of joy. When the adult laces his boots, switches on the light, takes off his clothes in evening, eats with knife and fork…, these performances, that all had to be toilsomely learnt, do not in the least disturb him in the thoughts in which he may just be engaged" [54].

The important point here is that with the accumulation of life experience, we learn to correctly "arrange" the objects in space in a perspective manner. This is based on the knowledge of both proportions and their change with distance (reachability), which was most deeply felt in childhood, as is mentioned by Schrödinger. Indeed, the most striking emotions of our earliest ("unconscious") period of life are associated with spatial perception. Perhaps, that is why we as adults do not remember this and take the space for granted as a fact of life completely coinciding with reality. Only watching an infant can one realizes the strong emotions induced by the "reachability" of something, when any object stirs interest but is not always within reach. A child, with time, starts moving and creeping, thereby expanding the range of "reachable". Spatial perception in the norm is inseparable from visual impressions. This is the particular stage when the links of "bigger–smaller" and "closer–farther" types emerge in a child. The very reachability is experienced as an obligatory sequence of efforts to be made in order to touch or take a desirable object. Moreover, a sequence of visual patterns, the frames associated with observation of changes in the angular dimensions of the "target", is associated with this sequence of efforts. Fig. (**3.1**) shows how spatial illusions emerge.

If we make a mistake in assessing the "reachability–rigidity", the estimated sizes of the objects are distorted. Numerous optical illusions utilize this effect, namely, the incorrect assessment of rigidity. The so-called Ames distorted room is a good example (Fig. **3.2**).

It is remarkable that both Poincare and Schrödinger posed the question of whether it is possible to represent reality in a different manner by omitting the conventional notions, such as objects, space, and time. "…The great thing was to form the idea that this one thing—mind or world—may well be capable of other forms of appearance that we cannot grasp and that do not imply the notions of space and time. This means an imposing liberation from our inveterate prejudice. There probably are other orders of appearance than the space–time-like" [54]. Hopefully, we "will see" different content in the directly observed picture at a certain stage in development of our consciousness.

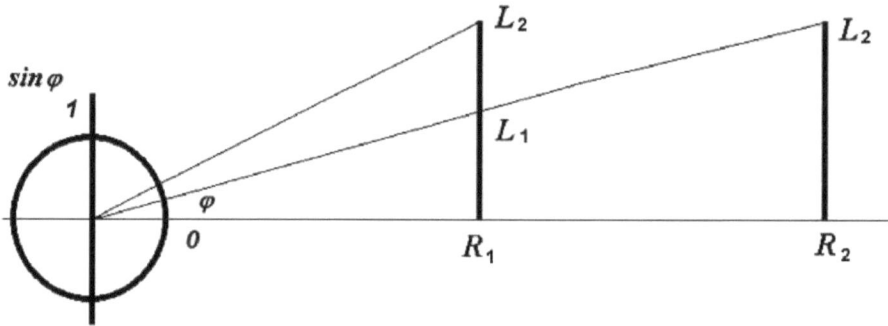

Fig. (3.1). An erroneous perception of the distance to an object, R (or its size, L) results in distorted estimate of its size (or the distance to this object).

Fig. (3.2). Perception of the Ames room and its actual design (from the internet).

Thus, the selected psychosemantic method for the description of objects (representation of objects on the mental map) leads to a complete agreement with the physical concept of reality. Undoubtedly, the obtained relations are objective and common for the mental maps of all *Homo sapiens* representatives. However, they refer to *a mental method of reality representation rather than to this reality itself.* An example is some analogies to the paradoxes in cartography, in particular, when a practical need forces us to map the globe in flat Euclidean projections, as in the case of topographic maps. No paradoxes emerged until our needs were confined to small areas, which determined the specific qualities and patterns of our perception. However, it is evidently unfeasible to construct an integrated "flat" map of the entire world without distortions, which increase with approaching the special points (typically, the Earth's poles). Naturally, the topological and metric distortions on a flat map can be expressed as *a precise "objective" law;* however, it is related *to the method of reality representation* (on a plane or a sphere) rather than to the very "objective reality".

Summing up, we have several additional comments on how a sign is formed and time is represented in an "individual consciousness". We will every now and then revert to this issue since the notion of time is a key point in science, which has given rise to numerous fallacies having nothing to do with reality.

We describe a perceived process, $S(t)$, as a certain function. In order to construct a **sign** (or a **marker**) for a *process,* it should be "fold-up" in the time domain. Consider spectral transformations as an example. Each function can be either strictly harmonic or nonharmonic. In the latter case, the function is describable with a certain analytical expression, particularly a Fourier series or integral, *i.e.*, it can be represented as a sum of *harmonics.*[22] Then, the amplitude is also a certain sum of coherent *elementary standard harmonics, i.e.*, it is discrete. It also follows that we may not use the concepts of *instantaneous* amplitude or frequency *that changes in time* since they automatically *generate other harmonics.* This approach brings about several problems with suspiciously numerous analogies to the problems in quantum physics.

CONCLUSION

In psychology, each individual possesses and is described by his/her unique set of qualities and their intensities: strength, emotional balance, flexibility of nervous processes, temperament, behavioral pattern, and so on. Actually, an individual is *a particular observer's reference frame* in perceiving and assessing the "external" processes. All external phenomena will be perceived differently *in different observer's reference frames* because of individual distinctions.[23] Note that the psychological observer's reference frame is determined by the degree to which *psychological qualities* are manifested, similar to the physical observer's reference frame and the corresponding *physical qualities* (velocity, acceleration, *etc.*).

It is high time now to consider the distinctions between the notions of "reference frame" and "observer's reference frame" in more detail. In the model by Poincare–Lorentz, the transformation of coordinates with a certain velocity V as the argument corresponds to the spacetime of each reference frame. In this case, an own spacetime is associated with each observer, significantly different from the variant of Galileo–Newton. Now a real process considered by different observers is described with its own model by each observer, and these models are similar but in no way identical [55]. All of this is a direct consequence of the prohibition, *i.e.*, the unfeasibility of determining the absolute velocity of an inertial reference frame.

As it has emerged, we know about reality only what "is given to us in sensations". It is necessary to emphasize the consistency of development of our scientific concepts from the model of reality with one subject by Galileo–Newton through the multisubject model by Poincare–Lorentz to eventual recognition of the fact that we may speak only about our own individual "worldview". In communication, we have to agree on our "mental maps", *i.e.*, our own worldviews, since that is all that we actually have.

Thus, the reference frame in physics is an observer with the linked system of coordinates and a clock. Unlike the reference frame (in general), the observer's reference frame can be defined as an observer who, among other things, has his/her individual mental map, making such a system unique. This difference emphasizes the fact that different observers' reference frames are not identical but rather coherently similar; in addition, this imposes physical limitations on the movement of observer in the spacetime model of reality.

In physics, we have learned how to describe a phenomenon in terms of a certain mathematical construction; in a similar manner, it is possible to propose a model of our unconscious that would explain the principles underlying its function. The most important features associated with the algorithms involved in constructing the mental map that reflects the reality we perceive are considered above. Correspondingly, the next chapter dwells on the construction of an object-based space that describes the "work" of our unconscious as an example.

NOTES

[1] This part of the work was supported by the Russian Foundation for Basic Research (project no. 19-013-00603a).

[2] Associated with the sphere of needs of an individual.

[3] That is the signals associated with satisfaction of the relevant needs of an individual and orienting the individual in his/her search for "external means" necessary to compensate for disturbances of "internal homeostasis" (*meaningful*

for this individual).

[4] To be more exact, "our" genetically determined mechanisms associated with the function of our internal sphere.

[5] Evidently, neither a spatial point nor a time point (instant) can be perceived; they are a pure abstraction.

[6] The Nyquist frequency.

[7] A particular sensation is already differentiated from the other sensations in its quality and intensity and is a *sign marking* a certain ith process and its intensity as an integral characteristic of the function $s_i(t)$ over time Δt. The quanta of perception time, Δt, are automatically regulated by the adaptive mechanisms of tuning for the significant signals, for example, tuning for the informative frequency of sound.

[8] The realization of the meaning (or perception of the sign) is a *qualitative* change and *an actually instantaneous* event. Unlike the *process,* the meaning does not exist in spacetime. Obviously, a piece of information stands behind a sign; however, physicists do not search for signs, for example, on an electron, to understand "where" and "how" the rules of its behavior in electromagnetic fields or in its interaction with other objects are specified (actually, this is how a "psychophysical problem" is formulated). It is evident that this "knowledge" belongs to the overall Universe as the totality and is unrepresentable in a local manner (is not contained in a spatially localized object). The laws of physics do not "belong" *to individual electrons* and are not "spelled out" as behavioral programs in "the brains" or some other mechanisms but rather *implement their content as processes* via these object-based forms, since the representation of *content* in spacetime is unthinkable in any variant other than via *processes* of information (patterns) translation.

[9] Space is a logically conceivable form (or structure) *that serves as the environment* where the other forms or particular constructs are implemented. The current mathematics defines space as a set of certain objects referred to as its points, with objects represented by any geometric configurations, functions, or states. Regarding their set as a space, only the qualities of their totality that are accepted by definition are taken into account. The relations between sets of points

determine the space geometry. The main qualities of these relations are expressed in the corresponding axioms. Actually, space determines the possibility (in terms of the corresponding qualities) to construct particular objects and to specify particular processes and states.

[10] Depending on the psychic state of an individual.

[11] Signal as an informative part of the process it reflects that has the meaning for *individual and is significant to individual.*

[12] This aspect of the origin of speech is widely described in the relevant literature (for example, [50]).

[13] The signs of this type are valid and can be used for description of objects as points by denoting their *coordinates* in a certain *semantic subspace,* i.e., can be used as a "second signal" modeling of the "first signal" *mental map* of an individual [49]. Naturally, these points can be further *named* as A, B, C, etc. However, the metric traits and a "coordinate" nature are still implicitly present in their values, for example, the words *"apple"* and *"pear"* for us are *semantically* "closer" to one another than the words *"apple"* and *"camel"*. Note that a coordinate-based representation of a *named object* on a mental map is frequently non-evident since it is initially constructed in a sign-based representation of *the first signal system* but is designated *in the second signal system* (according to Pavlov). However, we are forced when *modeling* a mental map in the semantic space to implement representation of the named objects *in an explicit form* making evident the "semiotic roots" of the name.

[14] The boundary implies specification of a quantitative threshold in the frame of a certain quality. A measure is necessary to determine a quantity. Even when comparing abstract numbers, we still imply that this comparison takes place within the same measure: indeed, we do not regard 3 hours smaller than 15 minutes and, the more so, than 15 hectares. The boundaries of an object are determined by *quantitative* degree of the intensity of its *qualities* (i.e., within the limits of certain qualities).

[15] For example, the expressions "morning star" and "evening star", defining the same planet, Venus, are different in their meaning.

[16] For example, dress can be not only the mean for satisfying a physical need in temperature comfort, but also a way of elevating the social prestige, sign of high social status, and so on, which is determined by different qualities of dress unequal in satisfying different needs.

[17] For example, the rating of objects in the class "dress" (*shirt, sweater, and coat*) with the invariance in their meanings will differ in the case of *hot* and *cold* weather.

[18] Usually, psychosemantics regards the semantic space as the space of primary qualities that describe the totality of objects after its factorization. As a result, this gives sets of linked descriptors interpretable as factors of social attitudes. In this approach, different meanings of an object are uncontrollably "averaged"; correspondingly, the object ceases to match its denotation. In addition, individual psychosemantic studies become actually incomparable since each is implemented in its own factor space [51, 52].

[19] It is still difficult to imagine that the definition of objects of different types requires different principles underlying construction of mental spaces. We will not discuss here the absurdity of this assumption and confine ourselves to Occam's principle that *entities should not be unnecessarily multiplied.*

[20] The sum of squares is taken only to eliminate the sign of projections since we are interested here in the direction alone.

[21] Indeed, the *mass* in physics is by definition the *resistance* of the property, such as *velocity,* to the action of the force directed to change it. As for rigidity, this is actually the resistance to changes.

[22] Most likely, the spectral representations play a significant role in our sensations, not least because the intensity of all stimuli is encoded in neurons in a frequency-dependent manner.

[23] Depending on the degree to which our own qualities (anxiety, communicability, working capacity, *etc.*) are manifested, we will assess the same qualities in other individuals. For example, Russians believe that tranquility is a national characteristic of the Finns and hot temper, of Georgians. That is why even a choleric Finn will look as phlegmatic for a Russian.

CHAPTER 4

An Object-Based Model in Physics

Мы жизни длим мгновенья, когда приходит вдохновенье,
И сокращаем их опять, когда нас некому понять.

When inspiration fills the mind, we stretch the moments of life. Alas,
We shrink them back when there's no one who understands us.
Midnight thoughts

Abstract: A topological model of the object-based space is represented as a bundle allowing for a complete description of an object by specifying the intensity and rigidity of its properties, and the object itself is represented as a resultant vector. As a case study, an object-based space is constructed that makes it possible to obtain the relations of the special theory of relativity as the conservation laws of informational content without using the hypothesis on the existence of space and time.

Keywords: Classical and imaginary time, Mental space, Object-based space, Quantum mechanics, Special theory of relativity.

INTRODUCTION

Not only does the necessity to respond to any change in the situation in a real-time mode underlies the existence in this world, but rather the need to respond feedforwardly since our signal system is time-lagged. In turn, this suggests that we have a model of reality with prognostic properties. This is an unconscious model, which, perhaps, explains that it is not very "advanced"; nonetheless, this model has emerged to be sufficient for our species to survive. Interestingly, there have been no serious enough attempts to mathematically describe this model.

It is clear that the central part of our model of reality is the "perception" of the surrounding world as an object-based entity, which means that it can be decomposed into constituent objects. Before such global decomposition, the objects themselves must be formed *via* assembling a certain set of characters and properties. Let us give a mathematical description of such assembly in the psychosemantic paradigm, which eventually leads to the construction of an "object-based space".

Sergey P. Suprun, Anatoly P. Suprun & Victor F. Petrenko

Construction of an Object-Based Space

We need to formulate the necessary requirements for such construction. Evidently, the object in this space must be specified as a set of certain properties in the form of a resultant vector, the components of which are qualities expressed by the corresponding axes. Note that the property in psychosemantics is not a simple value but is rather characterized by a quality (modality), its intensity, and rigidity (resistance to change), which must be expressed in a mathematical description. In addition, an object-based space "must" follow the conservation laws in the interaction of objects or a change in the observer's reference frame.

It is possible to determine some specific features of the unconscious semantic space by analyzing the simplest patterns of our perception. For example, when pooling two similar cups of coffee (considering the content of the cups as an object) placed on a table in a moving train, **we will say without a second thought** that the taste, color, and velocity relative to the table or earth of this **new** object have not changed [6]. However, the mass, volume, and other derived characteristics have doubled. Thus, rigidity (which reflects the resistance of quality to a change in its intensity, for example, mass with respect to velocity), is summed up similar to any value representable on the axis of real numbers. However, other qualities are not additive and require some other kind of description. Their intensity can be identified with the value of an angle, for example, in a unit circle. The situation can be explained in the following way. If you ask about the way to points A and B situated in a straight line from you, you will be informed about the direction and the distances to the first and the second points. The way you have to cover is the sum of the distances from the point where you are, to the first point (A) and from the first to the second point (B). However, you do not need to sum the direction (an ingle in a certain system of coordinates).

Now, imagine an object with a single property—a rest mass m_0—represented by a vector running from the center of a unit circle (an orthonormal basis) at the angle, the sine of which corresponds to the intensity of the property ranging from −1 to 1. As mentioned earlier, a "valid" property allows for normalization. In particular, normalization is performed by the classification of objects, for example, into solids, liquids, and gases. In this case, the characteristic, such as temperature, which is defined in physics on a semi-infinite scale, has the meaning for liquids only in the interval from the melting point to boiling point. However, any finite interval is simply converted into a unit interval by introducing a normalizing coefficient, which allows us to easily reply to the question of what is the temperature of the pooled liquids, at least on a qualitative level.

In a mathematical representation of the mental map, the most evident and natural is mapping of the **mental space** of the "first signal system" into the *vectorspace of the properties* of the "second signal system", which, in fact, is implemented in an implicit or an explicit manner by linguists in the case of a direct[1] (absolute) definition of objects *via* their properties, $\mathbf{O} \rightarrow O = O(q_1, q_2, ..., q_n)$. Here, object **O** of the mental space is mapped into vector $\vec{O}$ in the space of properties $\{q_1, q_2, ..., q_n\}$. It is easy to show the inadequacy of this mapping using the above example with the cups of coffee: we pooled two portions of coffee, O'_1 and O''_2. In a vector summation of these objects, the coordinates are added, that is, if

$\mathbf{O'} \rightarrow \vec{O}';\ \ \mathbf{O''} \rightarrow \vec{O}''$ and $\mathbf{O} = \mathbf{O'_1} \bigcup \mathbf{O''_1} \rightarrow \vec{O}_2$, than $\vec{O} = \vec{O}'_1 + \vec{O}''_1 = O_2(q_1' + q_1'', q_2' + q_2'', ..., q_i' + q_i'')$. However, the taste ($q_1$), smell ($q_2$), and velocity relative to the earth, in this case, have not changed, $q_1' = q_1'' = q_1$; $q_2' = q_2'' = q_2$; and $q_3' = q_3'' = q_3$, although the mass of the object (q_m) and the associated characteristics (volume, weight, *etc.*) are actually summed. In our representation, such a situation is possible only if we concurrently use both the *angular* and *linear* coordinates. Actually, the difference is determined by the possibility to introduce a limited normalized basis in one case and only the norm of space in the other. It is evident that the vector's angular coordinates (q_1, q_2, q_3), which determine *its direction,* do not change if its length is doubled. However, its length, which we have to relate to *mass m,* will, in fact, increase twofold.

Consequently, in order to make the mapping $\mathbf{O} = \mathbf{O'_1} \bigcup \mathbf{O''_1} \rightarrow \vec{O}_2$ adequate to the mental map even when describing *one property* q_1, we need *not a one-dimensional space* but *a plane*[2]. Let $q_i \rightarrow V_i = C_{(i)} \sin\varphi_i$ (where $C_{(i)}$ is a certain constant that determines the measure of or the scale of variation in this property); then the object is represented as

$$\vec{O}_i = \left\{ |\vec{O}_i| \cdot \sin\varphi_i; |\vec{O}_i| \cdot \cos\varphi_i \right\} = \{P_i; U_i\}.$$

Since an angle can be specified in a space with at least two dimensions, $|\vec{O}_i|$ is the length of projection of vector $\vec{O}$, which describes the object in all its properties, onto the "plane of its *i*th property", $\vec{e}_{Pi} \times \vec{e}_{Mi}$.

It is possible to use the designations of properties accepted in physics when describing the objects:

$$q_i \rightarrow V_i = C_{(i)} \sin\varphi_i : \sin\varphi_i = V_i / C_{(i)} \ ; \ \cos\varphi_i = \sqrt{1 - \sin^2\varphi_i} = \sqrt{1 - V_i^2 / C_{(i)}^2},$$

, consequently, $|\vec{O}_i| = \dfrac{U_i}{\cos\varphi_i} = \dfrac{U_i}{\sqrt{1-\sin^2\varphi_i}} = \dfrac{U_i}{\sqrt{1-V_i^2/C_{(i)}^2}}$ (Fig. **4.1**).

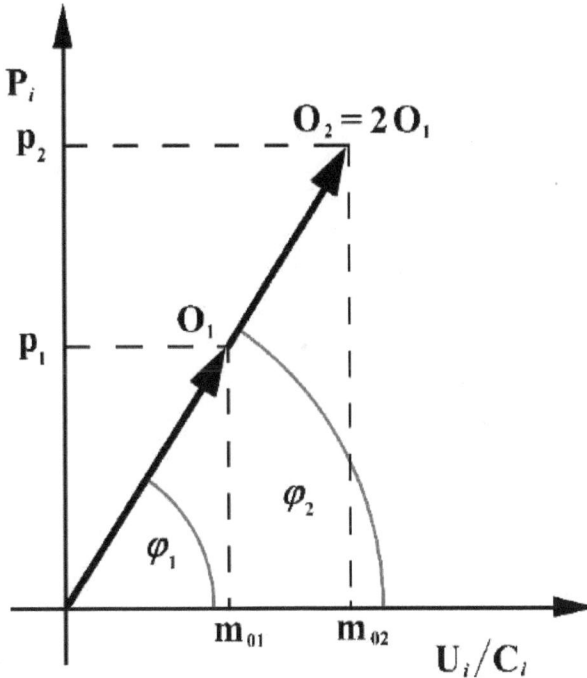

Fig. (4.1). Representation of an object on the mental map $(\mathbf{O} = \mathbf{O}_1' \cup \mathbf{O}_1'' \to \vec{O}_2)$.

Then,

$$P_i = |\vec{O}_i|\sin\varphi_i = \dfrac{U_i}{\sqrt{1-V_i^2/C_{(i)}^2}}\cdot\dfrac{V_i}{C_{(i)}} = \dfrac{U_i/C_{(i)}}{\sqrt{1-V_i^2/C_{(i)}^2}}V_i = \dfrac{m_{0(i)}}{\sqrt{1-V_i^2/C_{(i)}^2}}V_i = m_{(i)}\cdot V_i$$

which coincides with a relativistic *momentum-based description of a physical object* (in the case of a single property V). Here, m_0 is the rest mass; m, gross mass; and P, momentum. In this representation, it is easy to make a mistake and to regard *the rigidity of a quality* of a certain object as *an individual independent quality*, which actually takes place in psychology (and, besides, in some physical theories, for example, in the theory of gravitation). As has actually emerged, physicists, when switching from velocities to momentums, implicitly make the transition from an incomplete *space of qualities to the semantic space* adequate to the mental map, which is the only one able to retain the semantic components of the description of the interacting objects[3].

For the object with a determined rest mass m_0, it is always possible to construct the section of a cone perpendicular to its axis at the point corresponding to the m_0 value. The plane of this section houses the circle with radius $\{O_1, B_1\}$, corresponding to m_0 for an angle of 90° at the vertex of the cone, which is the basis separately, as shown in Fig. (**4.3**).

In principle, an object-based space—the space that permits a complete description of an object *via* specifying the intensity and rigidity of its properties—is conceivable as a bundle or, more precisely, as a *direct product* [56–59] of an orthonormal space of properties and a linear space, the axis of real numbers reflecting the mass (in a general case, rigidity). The simplest case of *a direct product,* which is frequently used as an example, is a circle (*base space*) and the real axis (*layer*) normally oriented relative to *the base.* In this case, the result will be a cylindrical surface as the product of two spaces. The variant of our interest is somewhat more complex. Fig. (**4.2**) shows the *layer, i.e.,* the real axis (m_0 is the positive part from zero and farther), and the *base space,* a section of a cone, the plane of which houses an orthonormal basis (unit circle). The axis corresponds to the mass of an object, while the orthonormal basis reflects the relative intensity of a certain single property.

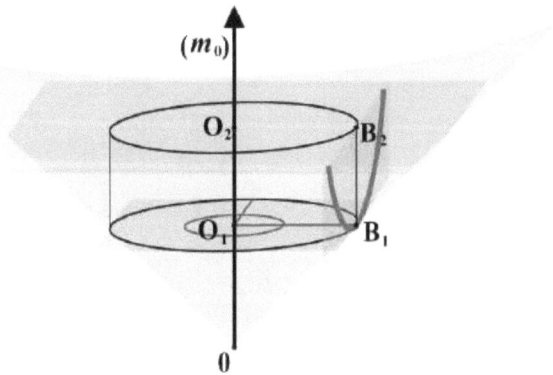

Fig. (4.2). The object-based space as a bundle.

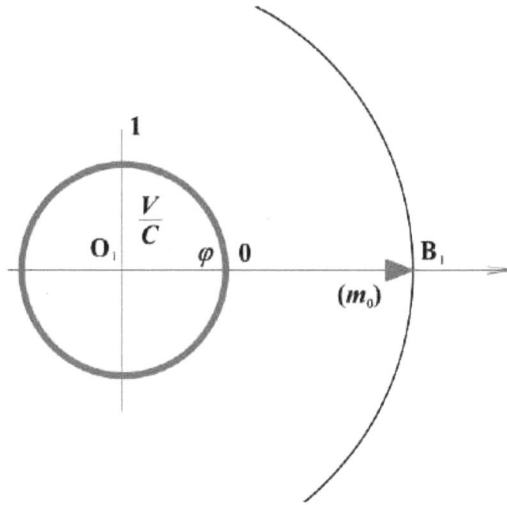

Fig. (4.3). A plain section of the object-based space.

If we consider velocity V as such a property (normalization by the velocity of light, C, allows for the introduction of the unit basis; sine of the angle amounting to 0 and 1 is shown on the circle), then the intensity of this property is described as $\sin\varphi = V/C$. We introduced such a designation for the mapped intervals in order to "fasten" to the physical interpretation of the described picture; however, any properties may be used.

It is necessary to mention the difference between the gravitational and inertial masses. This difference consists in that the mass as rigidity relative to velocity, *i.e.*, as a resistance to a change in this property of object, is perceived directly by the observer in the own reference frame. This is what corresponds to the inertial mass in an object-based model. As for gravitational mass, most likely, this is a property of the "systems" model since it depends on the mutual arrangement and properties of the parts of modeled reality as a system. However, these two masses can quantitatively coincide for the observer who uses an object-based model and who is situated inside a gravitational system.

In this section (Fig. **4.3**), the object at rest is represented by vector $\{O_1, B_1\}$. Now we make the additional constructions shown in Fig. (**4.2**): the cylinder with the base of the section in question (Fig. **4.3**) and one more section of the cone by the plane that touches the cylinder's surface at point B_1. The last section will be a hyperbola with the vertex at this point (Fig. **4.3**).

Now let us describe the algorithm that makes it possible to consider the changes in the properties of an object relative to its rest state, when the intensity of a property is zero. In our model, the end of vector (O_1, B_1) (Fig. **4.4**) with an increase in the intensity of the property expressed by angle φ rises from the "rest" plane and moves along the hyperbola. Note that the difference between vectors $(\mathbf{O_1}, \mathbf{C_2})$ and $(\mathbf{C_2}, \mathbf{B_1})$ or their projections $(\mathbf{O_1}, \mathbf{C_1})$ and $(\mathbf{C_1}, \mathbf{B_1})$ on the "rest" plane is always equal to the constant value of $(\mathbf{O_1}, \mathbf{B_1})$, *i.e.*, m_0.

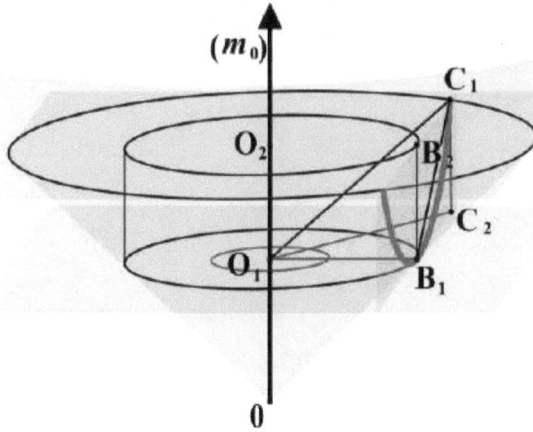

Fig. (4.4). Change in the intensity of the property of an object.

Consider a new section specified by the plane $\{\mathbf{O_2}, \mathbf{B_2}, \mathbf{C_2}\}$ and the corresponding increase in the intensity of the property. In principle, it may be regarded as a new object, since its properties differ from those of the initial object at rest. However, we focus on an assumed situation that permits a quantitative change in the intensity of properties but without altering the overall entity. Then the projection of this section onto the initial rest plane, as is shown in Fig. (4.5), allows for the construction of vector $(\mathbf{O_1}, \mathbf{C_1})$ (also designated μ $E\cancel{\,}\mathcal{C}$, where μ is a coefficient with the dimensionality of mass), which is an integral characteristic of the object. This vector reflects both the degree of intensity of the V/C property and the initial "rest mass" of the part of reality that is selected as the object. It is necessary to emphasize that the object per se in such a construct is specified by vector $(\mathbf{O_1}, \mathbf{C_2})$, while its projection $(\mathbf{O_1}, \mathbf{C_1})$ reflects *the change in intensity of the property expressed via* the change in mass.

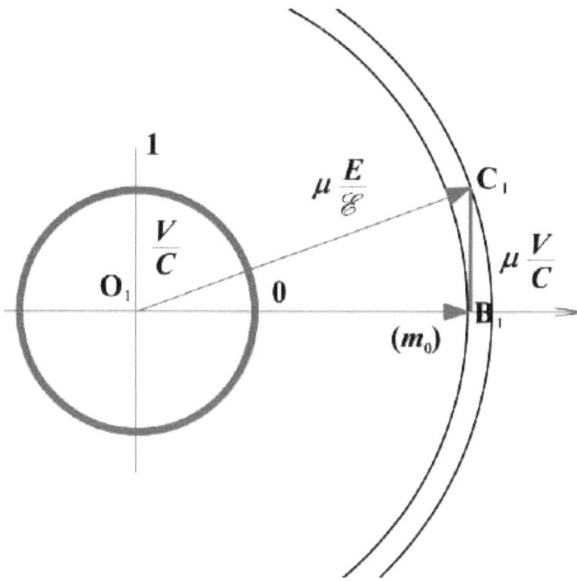

Fig. (4.5). The changes in intensity of the property of an object relative to the rest frame.

It is evident from Fig. **(4.5)** that $tg\varphi = \dfrac{\mu}{m_0}\dfrac{V}{C} = \dfrac{\sin\varphi}{\cos\varphi} \rightarrow \mu = \dfrac{m_0}{\sqrt{1-V^2/C^2}}$. Analogously, it is possible to obtain the dependences between the momentum and energy of object, which are not its properties but rather its characteristics in this case. However, note that their interconnection, as well as the laws of conservation in this interpretation, means that the part of reality we consider as an object is conserved. Here, the observed changes in the properties of object are assumed reversible. It follows from consideration of the right triangle $O_1\,B_1\,C_1$ that

$$\mu E/\mathscr{E} = m_0^2 + \left(\frac{m_0\,V/C}{\sqrt{1-V^2/C^2}}\right)^2$$, wherefrom we get the dependence between energy,

momentum, and the so-called "rest energy": assuming $p = \dfrac{m_0 V}{\sqrt{1-V^2/C^2}}$ and

$\mathscr{E} = \dfrac{m_0 C^2}{\sqrt{1-V^2/C^2}}$, we obtain

$$m_0^2 = E^2 - p^2 C^2 \qquad\qquad\qquad (1)$$

A family of hyperbolas, shown in Fig. (**4.6**), can be obtained for different rest masses m_0.

In the object-based space, trigonometric functions of angle $\varphi = \arcsin (V/C)$ relate to the set of quantities (m_0, E, and p). On the other hand, there is a unique correspondence between $p(E, m_0)$, which is a hyperbola, and φ values. As is evident from Fig. (**4.6**), when the mass of object tends to zero, the hyperbola tends to a linear dependence at lower V/C values. Thus, the transition to the so-called relativistic limit (a Dirac cone) is not connected with an increase in velocity, as is commonly believed, but rather with a decrease in mass of the considered object. Now we formally get the linear dependence $E=p\,C$ for the particle that has no rest mass, for example, photon, although in the limiting case, the model may fail to work[4]. Generally speaking, the cone with a "punctured" vertex is homeomorphic to a cylinder, which, as is noted above, is a *direct product* of the spaces corresponding to circle and axis.

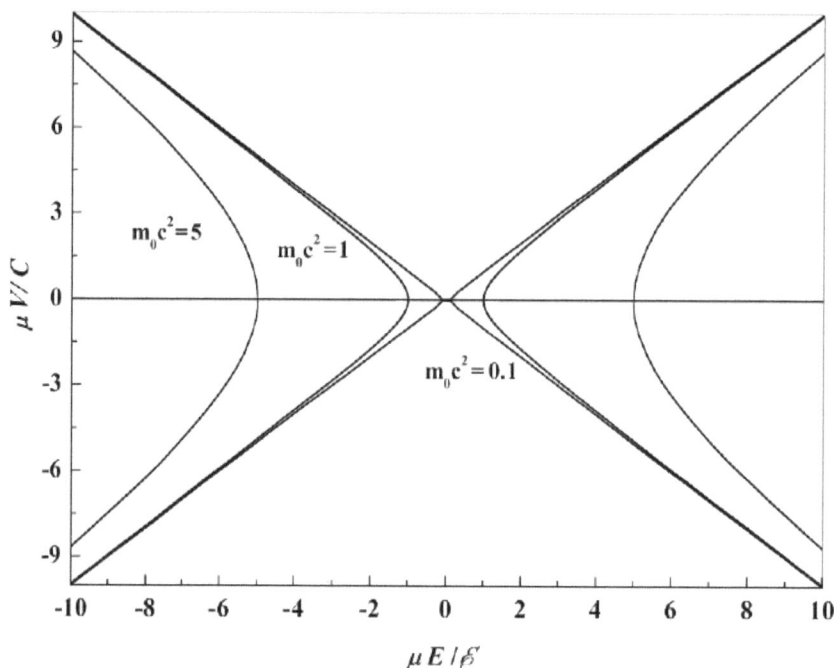

Fig. (4.6). The dependence $\mathscr{E}$ (E) at different values of the rest mass of an object.

The ratio $V/C = \sin\varphi$ in the orthonormal basis corresponds to the hyperbolic tangent for the dependence $\mu V/C(\mu E/\mathscr{E})$, as is evident from Fig. (**4.6**). In this case, the corresponding velocity V_S/C for a certain kinetic state can be resolved into the sum of the following form:

$$-\frac{V_S}{C} = \operatorname{th}\theta = \operatorname{th}\left(\theta_1 + \theta_2\right) = \frac{\operatorname{th}\theta_1 + \operatorname{th}\theta_2}{1 + \operatorname{th}\theta_1 \cdot \operatorname{th}\theta_2} = \frac{\left(V_1/C\right) + \left(V_2/C\right)}{1 + V_1 \cdot V_2/C^2} \rightarrow V_S = \frac{V_1 + V_2}{1 + V_1 \cdot V_2/C^2},$$

which coincides with the relativistic equation for addition of velocities

$$W = \frac{V_1 + V_2}{1 + V_1 \cdot V_2/C^2}.$$

This equation demonstrates the objective changes in the representation of the intensity of a property (for example, velocity) of one and the same object in the observer's consciousness depending on which share of observer's "experiences" related to this property the observer associates with *himself/herself* (V_1) and which, with *the object* (V_2).[5] In the STR, this corresponds to "the change in the observational frame of reference by observer" (virtual relocation of the observer from one inertial frame to another). It follows from the above said that our unconscious constructs within an object-based model can be embodied into strict mathematical forms. The point of such manipulations is, first and foremost, in that the problems initially matched to the actual reality rather than to our *ways of modeling this reality* become available for a realized and purposeful analysis.

The considered object-based space is applicable in physics, for example, when describing collisions of objects. Figs. (**4.7** - **4.9**) show a superposition of the bases of two spaces when the mass axes are directed towards one another. Then, Fig. (**4.8**) shows the transition to the center-of-mass system, which corresponds to the change in φ_i by the same angle in the direction towards one another analogously to the movement of the objects in a spacetime representation. In this case, all the laws of conservation for an elastic collision become "geometrically" evident (Fig. **4.9**).

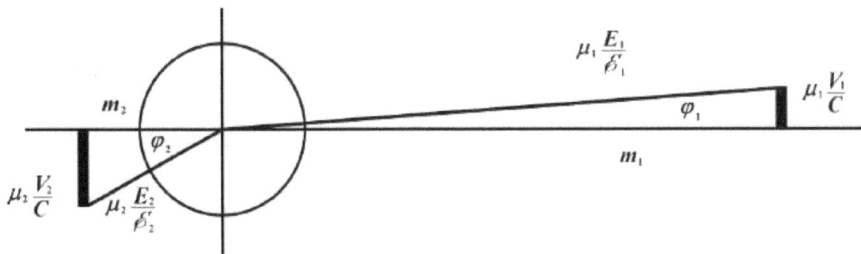

Fig. (4.7). Two objects moving towards one another.

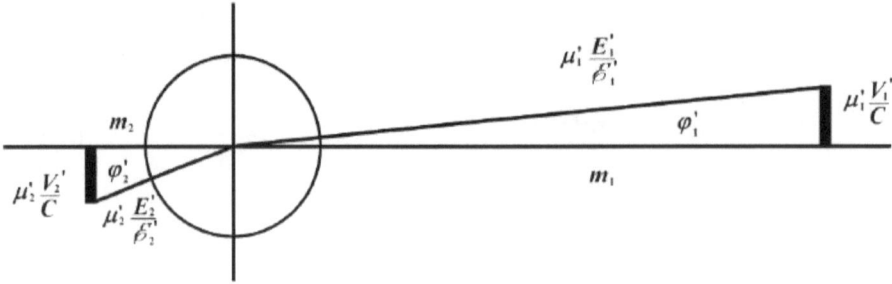

Fig. (4.8). Transition to a center-of-mass system.

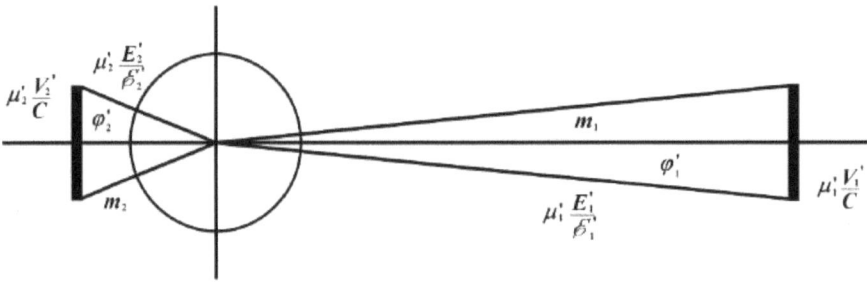

Fig. (4.9). The laws of conservation of mass, energy, and momentum during the collision.

As for the integral characteristic of an object, such as energy, the (O_1, C_2) vector length in Fig. (**4.4**) certainly differs from its projection (O_1, C_1), accepted in the STR as the total quantity of relativistic energy. The (O_1, C_2) value is computable with the help of a number of additional constructions and using the similarity of the triangles at the vertex of the cone. The difference in these energies can be tested experimentally and is of interest in our opinion. This means that, for example, the "mass defect" computed for these two cases will be different, as shown in Fig. (**4.10**), which is most pronounced at a velocity close to light.

Another interesting situation emerges when the number of considered properties of an object is increased, for example, by additionally considering the angular momentum (M). It is clear that the addition of one more circle with its center at point (O_1) in Fig. (**4.4**) located in the plane perpendicular to the mentioned section requires the construction of a four-dimensional cone. This is a more complex topological problem, which should be separately considered. Here, it is reasonable to make the following comments. Undoubtedly, the rotation, similar to the translational motion, is an independent degree of freedom. It is possible to normalize this property using the $\hbar/M$ ratio (more precisely, $\hbar/2M$), where $\hbar = h/2\pi$ is the Plank constant. Unlike the translational motion, the angular momentum in this variant takes the limit minimum value rather than the maximum one. At first glance, the remaining reasoning associated with this property is analogous to that for *V/C*. However, the attempt to introduce the angular momentum as an object's property gives an equivocal result. So, what is wrong here?

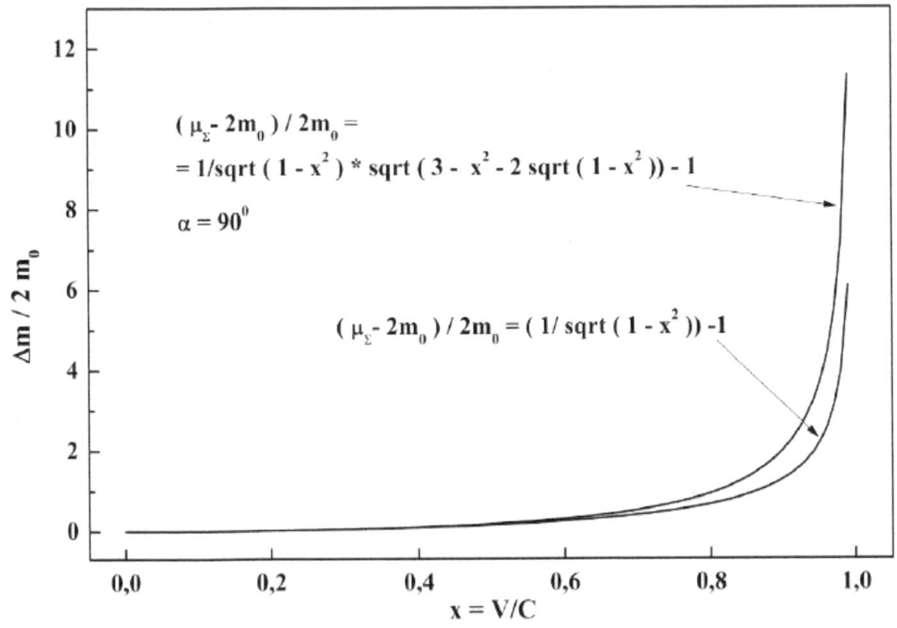
12
10
8
6
4
2
0
Δm / 2 m0
(μΣ - 2m0) / 2m0 =
= 1/sqrt (1 - x²) * sqrt (3 - x² - 2 sqrt (1 - x²)) - 1
α = 90°
(μΣ - 2m0) / 2m0 = (1/ sqrt (1 - x²)) -1
0,0
0,2
0,4
0,6
0,8
1,0
x = V/C

Fig. (4.10). The mass defect in fusion of two bodies each with rest masses *m*0 and equal energies *Eversus* V/C: α is the angle at the vertex of the cone (see Fig. **4.4**).

Generally speaking, there exists no uniform linear motion (even taking into account the space curvature in the model of general relativity theory) if we take into account the motion of the selected frame of reference within the planetary, astral, galactic, and other systems. Such an inertial frame of reference declared in the STR would be actually "timeless", since nothing in this system depends on time. Nonetheless, the translational motion is chosen as the "basis" for the kinetic processes. Even rotation in classical physics is actually reduced to consideration of "translational motion along a circumference". This is achieved *via* a spacetime[6] decomposition of the rotating system into objects and axis relative to which they are rotating (Fig. **4.11**). Note that not only continuous notions ***r*** and ***t*** emerge in a "well-grounded" manner, but also their spectral representations ***k*** and ***v*,** since the steady-state case is also "timeless" (and, generally speaking, is "spaceless" as well). Actually, such a description utilizes the same terminology as a wave-based description.

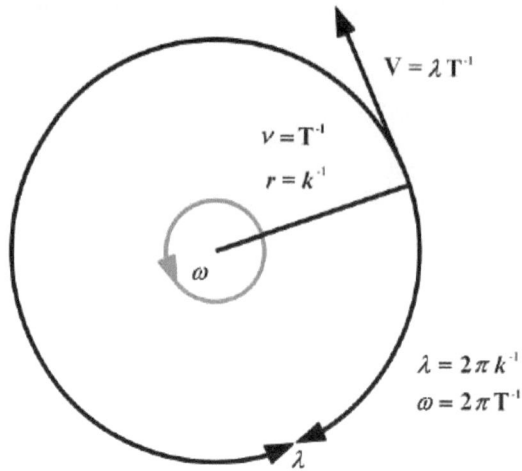

Fig. (4.11). The spacetime and spectral variables introduced when describing rotation.

The need to use the integral function, such as the moment of inertia, which reflects the distribution of mass in a system, demonstrates the systemic character of this property. In addition, mathematical notions are also somewhat different. The velocity in a translational motion is a vector: a true polar vector changes its sign when reflected by the plane perpendicular to it. Rotation is described by a pseudovector: a reflection of the rotation vector does not change the direction of motion because it takes place in the plane perpendicular to the axis of rotation. This picture becomes more intricate in quantum mechanics: as is known, the spin is unrecordable for a free electron. The intrinsic angular momentum, amounting to $\hbar/2$, is introduced to explain the results of the Stern–Gerlach experiment on splitting **the beam of neutral atoms** in a magnetic field. Thus, the electron in such a quantum system possesses a spin and lacks it otherwise. This problem is interesting since the "misunderstanding" appears already at the level of classical physics because of the absence of a systems-based description in science. Next chapters will consider the relevant consequences.

Thus, the classical part of physics can be considered in terms of an object-based model with the following applicability conditions:

1. Fulfillment of the criterion for object-based representation (observation of a certain value of a quality ascribed to an object is described in terms of classical probability with all the associated requirements, such as the commutativity of measurement operations, normalization condition, and so on) and
2. Fulfillment of the locality principle, which is a necessary and sufficient condition for applicability of spatiotemporal representation.

Certain "mental duality" that emerges when considering a quantum system results from the fact that although the used mathematical apparatus is completely adequate to the description of this system, the results are interpreted in terms of an object-based model. Currently, this is the only one existing human-understandable method at the level of unconscious. Recognition of the fact that we do not have and never have had any alternative to object-based modeling of reality (because the need in this option was absent in the course of the evolution) should be our point of departure when analyzing the existing situation. Moreover, this object-based model is a kind of "firmware" in our brains and its sign-based form has been further meaningfully developed by matching to the intuitive perception of reality. On the other hand, the mathematical apparatus chosen by the scientific community to describe the experimental results obtained when studying systems is independent of intuition (intuition-free). In this case, the object-based terminology and notions are forcedly used, whereas their initial meanings fail to match the situation. This is where all problems in quantum mechanics are rooted. Our confidence in the feasibility of infinite division and separation of a part of reality as a closed system provided that the principle of separability is fulfilled, has led us to a deadlock (note that a possible way for performing this operation is not specified; moreover, it does not exist).

We have no grounds to use an object-based model when considering a quantum system. The only way out is **to involve a spectral description as an alternative to an object-based one,** since the spectral description at least in a certain way meets the requirement that **the locality (separability) principle is not met.** In this case, the properties of a quantum system **before measurement** (the reality as an integral whole) are described using wave functions, while **after the measurement** with wave function collapse, the result of observation with a certain probability is translated into an object-based form. Since such an approach is never explicitly manifested, this brings about an irresolvable contradiction between these two stages of modeling. This is most likely determined by the incompatibility of the postulates (which have not yet been formulated in the scientific literature) of an object-based model and a system model.

An object-based decomposition of a quantum system makes it possible to assert with a certain probability that there is a certain "object" at certain time (Δt) that occupies a certain volume of space ($\Delta x \cdot \Delta y \cdot \Delta z$). This actually defines a spacetime "window" through which reality "is questioned" about the probability to detect a particular property. The equation $\Delta w \cdot \Delta t \geq 1$ is fulfilled for a spectral transformation; the multiplication by Plank constant $\hbar$ gives the known uncertainty principle $\Delta E \cdot \Delta t \geq \hbar$. Thus, the transition from an integral "quantum" representation of reality in a Hilbert space (beyond the consciousness) to an object-based spatiotemporal one in the consciousness in each time interval Δt is necessarily performed with a certain inaccuracy. This fundamentally distinguishes a quantum system from a "classical" object, a spatiotemporal existence of which is postulated *as the very reality*. The manipulation with a quantum system as an object fails to extract the full information about its properties. Actually, we incorporate the specific features of our perception of this reality into the physical reality (its "re-encoding").

Thus, the very method for description of *a quantum system* implies obtaining **spectral characteristics** as a result, whereas the properties of our interest are surely "object-based" (and we do not know any other kind of properties), namely, energy, momentum, and mass. The energy of a quantum system is primarily associated with the ability to resonate—to absorb and emit radiation with a certain wavelength. Momentum is the property ascribed to an element of a quantum system after its object-based decomposition, for example, to an electron in a potential energy well. Note that when representing the wave function of an electron in a potential energy well, the probability to detect the electron in different parts of the well is said to be proportional to the squared wave function and this is despite the fact that it is impossible to experimentally verify this statement. It is unfeasible to fix an electron or remove it from the system without a split of the system or loss of its properties. Another situation: spin as a phenomenon was determined in the Stern–Gerlach experiment using a beam of neutral silver atoms; however, this property (which is evidently systemic) is also ascribed to a free electron. However, it looks as if nobody is perplexed by the fact that the electron beam, when passing through the Stern–Gerlach device, behaves according to the Lorentz law in a completely different manner. Note that the fixed phrases used in the solid-state theory, such as "effective mass" or "quasi-momentum", clearly suggest the discrepancy between the used "sign" and its standard meaning, which reflects the conflict between the accepted object-based description and the systemic content of the considered phenomenon.

The Difference of "Time" in Classical and Quantum Physics

Вам пространством время станет, если нету сил шагать.
Позовет в дорогу память и часы раскрутит вспять.

> *Time will be space for you if you are tired to walk.*
> *The memory will prompt a journey and wind back the clock.*
> *Midnight thoughts*

An interesting aspect in the models of reality that we construct is the difference in the content of the term "time", a basic notion in physics. As was noticed long ago, the Schrödinger equation is derivable by replacing t with it (imaginary time) in the continuity equation (or heat conduction equation, or diffusion equation). Presumably, the "classical" (real) and "quantum" (imaginary) times are qualitatively different.

The former is used for "indexing" integral states of reality and can be thus specified as a parameter. As for the latter, this imaginary time is associated with the representation of these states as a kinetic spacetime model and as an attempt of their visual imaging in our consciousness. Naturally, any perception, for example, of an audio tone cannot take place *instantly* with respect to a "classical" time. However, the appearance (or realization) of an audio tome in the consciousness (determined in the unconscious over time interval Δt) *as something done* is instant. In the words of Alfred N. Whitehead, "...the temporal breadths of the immediate durations of sense-awareness are very indeterminate and dependent on the individual percipient... What we perceive as present is the vivid fringe of memory tinged with anticipation" [60]. Although the subconscious systems ("sensory physiological systems" or "mental mechanisms" according to an object-based interpretation, which is naturally used in all academic fields) translate the state into consciousness as a spacetime picture *over a certain interval* of "classical" time, we psychologically ascribe to the present any *currently lasting but yet uncompleted event* determined by a certain state. By the way, this is linguistically reflected in the verbal forms of various languages (for example, we say, "I read the paper", regarding *the overall time interval* of this event as the *present*).

From this standpoint, it is quite logical to return to an interval-based time scale in quantum mechanics. The uncertainty principle $\Delta E \times \Delta t \sim \hbar$, where ΔE is the interval (uncertainty) for the energy of state; Δt, time interval; and $\hbar$, Plank's constant. Note that quantum mechanics describes the situation of "present continuous", *i.e.*, *an uncompleted process* in which the final result (measurement) is yet unknown, although possible variants are known. Only an individual perception or the "measurement" by *an observer* leads to the so-called reduction of wave function, *completion of a certain stage of this process,* and its "transition" from an indefinite "quantum *present*" to a uniquely defined "classical *past*". Thus, the duration of "present" is *dynamic and does not collapse into a point.* Analogous phenomena are also observable in psychology by the example of the so-called Zeigarnik effect. Its essence is in that a person better remembers a piece of information associated with some unfinished processes rather than with the finished actions [61]. As we see it, this is explainable by the fact that until a situation remains in the domain of actual "quantum present", it is inaccessible to the "classical" mechanisms of forgetting. Korsakoff's syndrome, a kind of the amnestic syndrome preventing a person from making the current events of the present consistent with his/her own past [62], may also be associated with this phenomenon. Note that Korsakoff's patients have unimpaired memory for the current events, that is, unaccomplished actions.

As mentioned above, the quantum description is made with the help of the so-called ψ function, which reflects the potential variants of perception outcomes existing in the present with their probabilities. Evidently, all the variants of "the present" *should be consistent* with the already implemented unambiguous "past"; correspondingly, the Schrödinger equation for ψ function should be solved in the frame of well-defined *classical* initial and boundary conditions. Thus, it follows that the hope to create a purely quantum or a purely classical theory of Reality ("the theory of everything" as physicists use to say) is most likely doomed to failure, since the past of Reality and its present are *qualitatively different.* We just may not regard the time of "present" and the time of "past" as *the points of one and the same set that corresponds to a one-dimensional continuum.* In other words, our always multivariate models of the future reality cannot be "stitched" in a continuous manner with the model of the past since they "meaningfully" differ. Note that the *imaginary* (or "quantum") and *real* (or "classical") times on the Argand plane are mutually orthogonal (Fig. **4.12**).

In such a case, the "imaginary time" is directly connected with reality, whereas the "real time" is a merely dynamic representation of the content of a particular specific state of reality in the system of individual minds. (Philosophers frequently identify reality with the Absolute, but it should contain all possible kinds of the content (state) in *the actual present,* because the Absolute is always identical to itself, that is, timeless.) In this way, it is logical that the Wheeler–De Witt equation for the wave function of the Universe is independent of time (see https://arxiv.org/abs/hep-th/0211048). In this context, it is possible to draw an analogy of the reality to a collection of audio disks; moreover, the disks are selected in the "imaginary time" (beyond perception) and are played in the "real time". According to Stephen Hawking, "This might suggest that the so-called imaginary time is really the real time, and that what we call real time is just a figment of our imaginations. In real time, the universe has a beginning and an end at singularities that form a boundary to spacetime and at which the laws of science break down. But in imaginary time, there are no singularities or boundaries. So maybe what we call imaginary time is really more basic, and what we call real is just an idea that we invent to help us describe what we think the universe is like" [63].

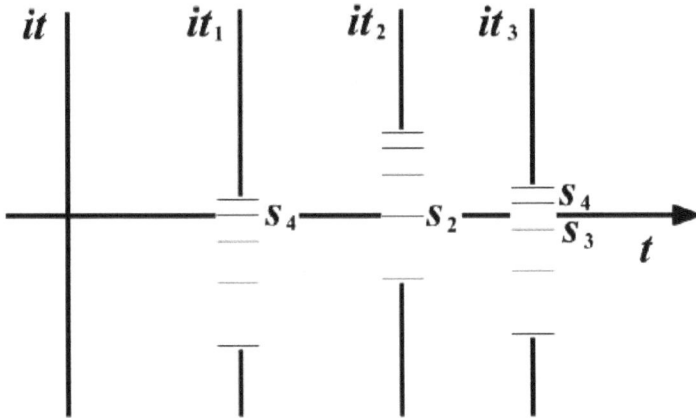

Fig. (4.12). Diagram of the change in the state of a quantum system (ordinate) *versus* the "classical" time (abscissa).

The "classical" real-time has no sense in a spectral description of quantum phenomena since the overall process there is represented as a *state,* being a whole entity (is represented in the entire space and time since the harmonics have neither time nor space limits).

Neuropsychology states that specialized "brain machinery" (V5 region) forms continuous motion of separate frames. The sequence of static frames, which are translation of certain signals through the spectral window, is converted by psychic mechanisms into a dynamic image. Thus, a continuous character of motion in our mind is an illusion created by the function of specialized systems (in an object-based representation, specific brain regions). In particular, persons with the V5 region injured by a cerebral stroke see the world as a sequence of stop-frames. They perceive all moving objects as static, changing their position in space "with a jerk". Thus, this illusion of continuousness of the perception flow gives rise to our experience of the "flow" of time.

In fact, our perception utilizes a frame-by-frame reading through the window of the quantum "present", which is orthogonal to our classical reality, and stitches these frames into a "linear" video sequence. As is mentioned above, the quantum-mechanical uncertainty principle (the energy in quantum mechanics is proportional to frequency) is, in this interpretation, obtainable from the well-known equation of the spectral theory: the product of the width of momentum spectrum (spectral window) by its duration is a constant, which determines the "sharpness" of the image.

The "classical" and "quantum" worlds may be interpreted from the standpoint of psychology as the consciousness and *that what is beyond its boundary, i.e.,* unconscious (or supraconsciousness?). Note that the means for representing these worlds are so different that our consciousness can only comprehend the "real" component of the "quantum subsystem". The world beyond our consciousness is an integral whole; however, it is translated into consciousness in a successive spacetime form.

The experiments on quantum teleportation of the so-called EPR pairs demonstrate that the *Universum as a whole* really exists. Penrose wrote about the EPR studies in quantum mechanics: "Despite their short of providing direct communication, the potential distant ("spooky") effect of quantum entanglement cannot be ignored. So long as these entanglements persist, one cannot, strictly speaking, consider any object in the universe as something on its own. In my own opinion, this situation in physical theory is far from satisfactory. There is no real explanation on the basis of standard theory of why, in practice, entanglements *can* be ignored. Why it is not necessary to consider that the universe is just one incredibly complicated quantum-entangled mess that bears no relationship to the classical-like world that we actually observe? In practice, it is the continual use of the procedure **R** that cuts the entanglements free, as is the case when my colleague and I make our measurements on the entangled atoms at the centers of our dodecahedra. The question arises: is this **R**-action a real physical process, so that the quantum entanglements are, in some sense, *actually* cut? Or is it all to be explained as just an illusion of some kind?" [7]. We believe that resolution of paradoxes in the current physics is possible only if we will separate as different systems *the reality itself* (as this "integrated whole" in a form of a set of states described in the Hilbert space) and its dynamic *representation in the consciousness* ("…the classical-like world that we actually observe" [7]).

Within a systems-level representation, the Universum may be described as an infinite set of all the possible states represented in the Hilbert space and translated through a system of "filters" into spacetime concepts of manifold subsystems (for example, "possible Universes") down to the subsystems of "individual consciousnesses". Psychology attributes this something wherefrom anything appears in an *individual mind* to unconscious, which, according to Jung [64] is "*collective*" (Wigner, a physicist, proposed similar ideas in his memoirs). A particular physical interpretation of a certain state of the Universum is implemented in a spacetime representation in individual consciousness at the interface of the so-called "collective unconscious" and "individual conscious-ness". This explains both the integrity of the Universe and the "subjective component" in the phenomenon of quantum reduction.

Naturally, a mechanistic coupling of the classical and quantum subsystems in the general theory of relativity becomes logically controversial similar to the controversial character of an object-based description in quantum mechanics. Here, we exclude from the physical paradigm both purely hypothetical and fundamentally improvable statements that tear the reality into the so-called "outward" and "inner" worlds, subject and object, materiality and ideality. Although a standard object-based representation of reality has played its part in the establishment and evolution of our consciousness type, nonetheless, this is neither the only nor the most universal metaphoric image of the Universum. According to Occam, multiplication of entities brings nothing else but logical contradictions (the psychophysical problem as such is self-contradictory and has no solution). The only thing that we can accept without contradictions is the existence of reality as an open set of linked evolutionarily developing subsystems with different types of representation of the subsets of the states of reality. Our consciousness is one of such specific subsystems. The universe of our individual consciousness is determined by the filters and the method used to translate the content of the Universum that had evolutionarily formed in a more general subsystem of reality.

CONCLUSION

In conclusion, it is possible to demonstrate the specific features of an object-based model in a spacetime representation and its incomplete adequacy to the observed reality by using the experimental foundations of quantum mechanics. We have considered the genesis of the notions of "space" and "time" and have demonstrated the feasibility of constructing an object-based space with its structure defined at the unconscious level.

It is necessary to emphasize that the STR is based on the hypotheses that space and time exist as fundamental physical entities independent of any observer and the possibility to introduce the notion, such as "inertial frame of reference", as well as on the Lorentz transformation laws for spacetime intervals in the assumption of a limited velocity for signal propagation. The above construction demonstrates that the STR conclusions can be evidently obtained through a formal consideration of the semantic space represented in an individual unconscious in the form of a mental map (in which the object-based space that describes the objects as such is part of the whole mental space that simulates reality). Such model of reality had formed during the evolution allowing an individual to orient in the flow of its sensations while satisfying various needs. In the further social development of *Homo sapiens,* this model has been correspondingly arranged and dressed in a scientific manner.

The relativistic laws of conservation are the laws of conservation of the meaning for what the observer specifies as the object and its content in the case of reversible change in the intensity of its properties or in the so-called transition from one frame of reference to another. It is essential that here we have not used the spacetime hypotheses (Einstein, in his letter to Bohr on March 03, 1947 wrote that physics must depict the reality in space and time and have no place for a mystic impact at a distance [10]. Everything has emerged far otherwise but has not received any interpretation complying with the experiment.) Most likely, Poincare was perfectly right when he insisted on a mental origin of these notions. The laws of transition of characteristics of objects in such an approach to modeling reality are the laws of the same type as, for example, the laws of perspective, which allow us to interpret under certain conditions a decrease in sizes of objects as their spatial distance from us.

As for the representation of the Universum as an open system, it is necessary to keep in mind two sides of the description, namely, an object-based (plurality) and subject-based (unity). Note here that neither a subject-based standpoint is equivalent to subjectivity nor an object-based one is equivalent to objectivity. In natural sciences, they do their best to retain an objective (that is, in its essence, an object-based) "depersonalized" language of its description, which excludes the subject (observer) from the paradigm. However, this fails to achieve the goal. The subject appears already in classical physics in the guise of an "observer", who is absolutely necessary to define the frame of reference. "Intangibility" of the observer appears in their ability to change their frame of reference instantly and without any energy expenditures. Note that in relativistic physics, the observer in this process quite *objectively* changes some physical characteristics of the perceived objects (according to the Lorentz transformations). As for quantum physics, the outcome of a physical experiment actually depends on the presence of an observer. A subject-based (integral) approach to the Universum and its object-based interpretation in an individual consciousness imply two different research methodologies, namely, a systems-based decomposition of the reality in the former case and a classical localized spacetime approach in the latter.

NOTES

[1] An indirect (relative) definition for an object is implemented when it is related to other objects with the help of metonymy and metaphor. An absolute definition is given *via* listing the properties of an object and their intensities, typically, in dictionaries.

[2] Since an angle can be specified only in a plane continuum.

[3] It is evident that physicists will not tackle prediction of the consequences of interaction between bodies without taking mass into account, unlike psychologists, who with ease approach such problems and explain their own mistakes by the complexity of the problem rather than the errors in problem definition.

[4] It is unclear how the mass of an object can be zero. As is known, both the numerator and denominator for and p in Eq. (1) go to zero in the case of photon: the rest mass is zero and the velocity is C.

[5] It is unclear in the STR how a change in *the frame of reference* differs from a change in the velocity *in the same frame*. On the other hand, as we see it, the change in the observer's reference frame should be accompanied by introduction of ω, an angular velocity. This is determined by that a direct manipulation with the object is possible only in an own frame and that the object possesses the property, such as velocity, only in this frame. In the remaining cases, we can observe only angular changes in the position of the object.

[6] Appearance of the spacetime terminology in description should immediately alert the reader in the perspective of what is has to do reality.

CHAPTER 5

Delayed-Choice Quantum Eraser in a Space-time Description

"Only a few find the way; some don't recognize it when they do; some don't ever want to."
Lewis Carroll
Alice's Adventures in Wonderland

Abstract: Analysis of the quantum physics experiments relying on the object-based concepts in space-time terms leads to irresolvable paradoxes. These paradoxes can be resolved by accepting that the wave function collapse, similar to time, matches the object-based model in consciousness that describes the result of observation and, moreover, the understanding of this means that the fact is fixed in the mental map.

Keywords: Locality principle, Object-based space, Quantum mechanics, Wave-particle duality.

INTRODUCTION

Currently, we may state that the understanding of quantum physics is absent. The results of most recent studies in this area look so mystic that evoke strong doubts on our knowledge about reality. There are several yet unsolved problems; however, they were regarded as philosophical ones irrelevant to practical activities until recently emerged to be the center of discussion of quantum experiments. What is space in physics? Why is time irreversible, although this does not follow the corresponding equations? And how it comes in this case that it is possible to "erase" the result of the measurement of a quantum state? What is consciousness, and how is it connected with the wave function collapse? Why can the quantum state be transferred at a velocity exceeding light? All these questions still have no answers, but as we see this, it is not right to avoid discussion of these issues.

The results of the experiments with EPR pairs were summed up by Nicolas Gisin in his recent monograph [65]. Gisin is one of the leading experts in this area; correspondingly, his view is most important for us in the further analysis.

Sergey P. Suprun, Anatoly P. Suprun & Victor F. Petrenko

Gisin precisely formulates the problems that have emerged in physics in connection with the existence of experimental foundations for quantum physics.

> "Quantum nonlocality took a long time to get accepted as a central concept of physics. Even today, many physicists reject the term 'nonlocal'.... And yet, from as early as 1935, Einstein and Schrödinger, among others, maintained that this aspect of quantum theory is its main feature."

> "...Today, violation of a Bell inequality is the very signature of the quantum world. But this is still a serious blow to our intuitions. Will the quantum technologies now under development make quantum physics and its nonlocality intuitive to us one day?"

> "...In particular, we still don't understand how to make this compatible with Einstein's theory of relativity. And nor do we understand the whole mathematical structure, nor the full potential of applications in information processing, nor—and this is perhaps the most surprising thing—the limits to nonlocality: why does quantum physics not allow more nonlocality?" [65].

Gisin writes in Conclusions, "Nobody knows why quantum physics is nonlocal" [65]. Here, he suggests searching for the explanation of the results by expanding the set of used terms: "But how can we explain or account for nonlocality? With our prequantum conceptual tools, it would be impossible, so we are compelled to extend our toolbox. One way is to speak about the nonlocal randomness produced by entangled objects" [65]. It is remarkable that Gisin is optimistic towards the possibility of solving the listed problems: "Personally, I have no doubt that nonlocality does, like relativity theory, generate some difficulties for our familiar concept of time, but from there to imagining a chronologically reversed causality which goes back in time seems a drastic step!" [65]. Thus, he (and not he alone) keeps the hope that the established scientific paradigm can be preserved without any considerable changes. Let us look at the essence of the problem so that our further conclusions do not seem radical but rather grounded.

EPR pairs are quantum *systems* represented in modern science as *a pair of objects*[1] displaying the properties *connected by the very way they have appeared* (Gisin uses the term "entangled objects" for a quantum system.) Hereinafter, we use the terminology and logic of the modern scientific paradigm "brought to a logical end". This means that we are forced to speak about what is understood but not reflected in the "academic" literature. Actually, we try to "bring to the light" the intuitive expectations, which are usually left unmentioned. Briefly, these unconscious concepts contain the following two points:

1. **Reality[2] exists independently of the observer in an object-based form** and
2. A space-time existence of objects is the reality itself.

Perhaps, these statements look rather straightforward, but they are formulated taking into account the call for making quantum physics intuitive: "Will the quantum technologies now under development make quantum physics and its nonlocality intuitive to us one day?"

Consider a case of the EPR pair formed in the annihilation of positronium. This quantum system comprises an electron and a positron, having a spin of ½, which is able to annihilate to give two photons. The process takes place only when the pair is in a singlet state, when the total spin is zero, *i.e.*, the spins are oppositely directed.[3] According to the conservation laws, the total energy, momentum, and spin of the formed particles, in our case, a pair of photons, will be equal to those in the initial system. Such states of emerged "objects" in quantum mechanics are referred to as "entangled". The cause underlying the correlation of their properties consists in the way they emerge. The above-considered example simply and evidently shows the emergence of correlation determined by conservation laws between the values of physical variables of two "objects" formed by annihilation.

Now we turn to the description of the experiments with photons and interpretation of their results from the standpoint of quantum mechanics, assuming a linearly polarized laser radiation. Having passed through an analyzer, the output radiation intensity depends on $\cos^2\varphi$ for the angle φ between the analyzer's axis and polarization of incident radiation. The same situation is observed in the case of a single-photon momentum, assuming a probability of $\cos^2\varphi$ to pass through the analyzer and a probability of $(1 - \cos^2\varphi)$ to be reflected. The sum of probabilities of these two mutually exclusive events is unity, and the probability of each event will be numerically determined by the basis, *i.e.*, the mutual orientation of the analyzer and polarization of incident radiation.

When considering the results of experiments[4] with EPR pairs [3, 34, 65], the following main statement can be formulated on their basis:

The necessary and sufficient condition that excludes a probabilistic description for the second measurement of an EPR pair is that this quantum system is measured using arbitrary but equal bases.

This means that the result of measurement of the second particle always becomes predictable with a probability equal to unity if this condition is met independently of the space-time interval between the measurements. The truth of this statement evidently follows from the type of wave function for the EPR pair. One would think there is no wonder: as mentioned above, the characteristics of these objects are correlated according to the conservation laws. In order to understand the paradoxicality of the implications following this statement in terms of the modern scientific paradigm, it is necessary to most explicitly formulate the hypotheses to be verified in these experiments.

Consider in detail the possible variants for an explanation of the experimental results from the standpoint of statements (**1**) and (**2**) regarding them as true.

First variant. Assume that emergence of an EPR pair in the process of a decay means that **the particles had actually existed before the observation started.**[5] This means that their properties also exist (particles or objects are identical to the set of their properties, *i.e.*, listing of the properties completely specifies the object [6]). Let the property, such as spin, have **a certain orientation in space at each time moment** for each particle (in the space-time that corresponds to the description of the object), for example, strictly orthogonal to one another (in a Hilbert space; Footnote 64). Then, the probability to obtain one of the two orientations when measuring the spin in a certain basis is $p_1 = \cos^2\varphi$ for the angle φ between the analyzer's axis and initial spin of the particle. The opposite result has a probability of $p_2 = (1 - \cos^2\varphi)$, which is also true for the measurement of the second particle. Therefore, the result of the second measurement in this variant is not predetermined ($p_2 \neq 1$)[6]. Thus, the coincidence (or orthogonality) of the results for an EPR pair when measuring on the same basis will be observed **not in all cases** but in accordance with the probability depending on the basis, which contradicts the experiment (the result excludes any probabilistic description). Consequently, **our assumption that the particles with a certain spin orientation exist in space at each time moment before the observation is wrong.**

Second variant. Any certain spin orientation does not exist before the first measurement of the system. In this case, we cannot state that the property, as well as the particle itself as an individual object, exists in space-time before the observation. There are no experimental grounds to state that they exist. Actually, the mathematical toolkit of quantum mechanics particularly suggests this when using a single wave function for the description of such a quantum system.

Assume that **spin is determined at the moment of the first measurement.**[7] However, this measurement, according to the conservation laws, determines the spin of the second particle as well, moreover, faster than the speed of light [65–67], as has been experimentally demonstrated. Correspondingly, **the principle of locality**, the essence of which is in that **a cause-and-effect relation can exist between the objects only in the case of a time-like interval** (within a cone of light), **is incorrect.** It is possible to avoid logical contradictions only assuming the statements of the second variant and its consequences as the truth.

Why is the principle of locality so important in modern science? The point is that all our reasoning is based on the assumption that the interaction between objects is transmitted in space (an object-based one) from point to point. "Until the advent of quantum physics, all the correlations predicted and observed in science have been explained by causal chains propagating contiguously from one point to the next, that is, by local arguments" [65]. Moreover, the carriers of interaction—electron, photon, graviton, and so on—are also declared as objects. However, common sense suggests that no material object can propagate at an infinite speed. "Entanglement is carried by quantum objects such as photons or electrons, and these objects propagate at finite speeds, less than or equal to the speed of light. In this sense, the notions of distances and space remain relevant, even though nonlocal randomness can show up in two arbitrarily widely separated places" [65]. The fact that teleportation transmits a quantum state and cannot transmit information [6] does not help much since this fails to answer the question on how it can be. It is necessary to pay attention once again to the point that an object-based model of reality with the corresponding space and time is used in all these cases.

Summing up of the results of experiments with EPR pairs suggests the following conclusions:

- **The absence of objects before observation in the classical space-time reproduction is experimentally proved** (more accurately, the object-based model fails in quantum mechanics) and
- **It is necessary to distinguish between reality and its object-based model in which space-time is either a "container" or a method for localizing the objects that specify its properties in our consciousness.**

Thus, statement (**1**) is wrong since reality is not object-based but certainly exists independently of an observer. In fact, it is an object-based model that does not exist without observer since this model is constructed by the observer's unconscious. As for statement (**2**), it is wrong because reality cannot be identified with its model. It is useless, which is regarded as a recognized fact, to search for the way out in terms of proper physics.[8] A broad view of the problem or, more precisely, analysis of the very foundations of the current science is necessary.

The most amazing thing in the current situation is that the solution was proposed even before this dead-end was encountered. Poincare in his works some hundred years ago highlighted that our space-time representation of reality was just *some mental schemes* [17–19]. His views on this issue are comprehensively analyzed in the previous chapters. Note that Schrödinger, in his last works, also came to the conclusion that the elimination of a subject from the model of reality made it objective only in our perspective.

"…But I set my heart here on discussing the other principle, that which I called objectivation."

"By this, I mean the thing that is also frequently called the 'hypothesis of the real world' around us. I maintain that it amounts to a certain simplification which we adopt in order to master the infinitely intricate problem of nature. Without being aware of it and without being rigorously systematic about it, **we exclude the Subject of Cognizance from the domain of nature** that we endeavour to understand. We step with our own person back into the part of an onlooker who does not belong to the world, **which by this very procedure becomes an objective world**" [54].

"So we are faced with the following remarkable situation. While the stuff from which our world picture is built is yielded exclusively from the sense organs as organs of the mind, so that **every man's world picture is and always remains a construct of his mind** and cannot be proved to have any other existence, yet the conscious mind itself remains a stranger within that construct, it has no living space in it, you can spot it nowhere in space" [54].

"The reason why our sentient, percipient and thinking ego is met nowhere within our scientific world picture can easily be indicated in seven words: because it is itself that world picture" [54].

The boldfaced phrases in the above statements by Schrödinger are the key to the understanding of the problems in physics and science in general. Each person individually constructs their own worldview, which is the product of individual consciousness; however, the illusion (and illusion only) of objectivity is achieved by exclusion of subject from the description.

This means that the only reality that exists for our consciousness is the one that is constructed by our own unconscious based on the "sensory experiences" (that is, the data of the first signal system, according to Pavlov). Correspondingly, it is useless to search for any other reality in it. The evolution has selected the most efficient processing algorithms, which have allowed us to feedforwardly respond to any situation in a real-time mode. Thus, the survival of a *Homo sapiens* individual and the species was guaranteed. Humans differ from the remaining species by that they have succeeded in developing a conscious sign model of a high level of abstraction, which makes it possible to predict the course of events based on the previous experience. Unfortunately, the established unconscious algorithms for primary data processing have emerged to be "tabooed" for analysis. As it turned out, the object-based (very simplified in its essence) scheme of the surrounding world proved to be the most efficient when solving the challenges of humans, first and foremost, the biological programs referred to as instincts. Yet this scheme has emerged to be maladjusted to research, and this is the heart of our problems. The fact is that Mother Nature could hardly set the goal of creating a being able "to construct a systems-level model of reality" focused on the understanding of quantum physics but rather "molded" an object-based model of Reality representation for some other purpose.

Interestingly, this representation appears to be very convenient for rapidly estimating an integrated state of a complex system. Once at a laboratory with the Institute of Biomedical Problems of the Russian Academy of Sciences, we were shown a device that allowed the psychophysiological state of a subject to be visually estimated by brain electroencephalography using 32 leads. A program that generated fractal landscapes on display according to several tens of parameters determined by each lead was used for this purpose. The type of landscape changed depending on the psychoemotional state of the individual and was easily assessable. This was an object-based spatial representation created utilizing a 32-dimensional wave process, which was initially in no way object-based. This representation undoubtedly "reflects" reality in an objective manner because of the objectivity of the visualization (transformation) program itself as well as the source of signals. However, two different factors correspond to this objectivity: while the first factor is determined by our unconscious, the second one in its meaning is comparable to the "external (physical) reality". Evidently, these two factors cannot be considered together as a physical reality because in this case we jumble up different entities.

In the further analysis of experimental papers in the area of quantum physics, we cannot but mention one of the central problems there, namely, the creation of a quantum computer, elaboration of the principles underlying its function, and the operation algorithms able to utilize all its capabilities in full. While searching for and selecting the prototype of a qubit register (qubits, or quantum bits, are the elements used for coding information and its processing), numerous quantum physics experiments have been conducted. All this has determined considerable advances in technology allowing for the design of a working model of a quantum computer and, concurrently, has revealed a number of problems associated with the insight into its functioning (according to Gisin, "And nor do we understand the whole mathematical structure, nor the full potential of applications in information processing…" [65]). Actually, the current situation ever more clearly demonstrates the inadequacy of our view of reality. As it has emerged, the ability to make a certain device does not mean the ability to use all its capabilities, which appears as a small number of applied problems resolvable using this device. On the other hand, the application of a quantum computer seems to have infinite prospects taking into account that it allows for a fundamentally different level in modeling of the mind. However, the estimation of the difficulties to be overcome in this way requires the relevant problems that have accumulated so far to be settled.

Xiao-song Ma *et al.* [3] published the most comprehensive review of the research into quantum systems with the help of the algorithms, such as quantum eraser and delayed choice, with a detailed list of relevant references. Now, let us analyze the results and conclusions of this review to clarify the level of understanding of the quantum problems. The main intrigue, resolved using the experiments exquisite in their design and performance, is the following: if a quantum system is able to behave as a wave or a particle, when the particular pattern of behavior is chosen and which conditions determine this choice.[9]

The behavior in terms of wave–particle dualism is tested relative to the phenomenon of interference since interference is a consequence of the property, such as the superposition of states of the tested system. Find below several quotations from the above-mentioned review [3], which clearly illustrate the position of its authors as well as the currently existing ideas in quantum physics.

"In Young-type double-slit experiments, every quantum system is at one point in time in an equal-weight superposition of being at the left and right slits. When detectors are placed directly at the slits, the system is found only at one of the slits, reflecting its particle character. At which slit an individual system is found is completely random. If, however, detectors are not placed at the slits but at a larger distance, the superposition state will evolve into a state which gives rise to an interference pattern, reflecting the wave character of the system" [3].

Note that the description of a quantum system is conceived in the categories of space-time, which reflects the perception of the very reality only in the frame of an object-based model. The authors unambiguously relate the presence of a superposition state of a quantum system to the alternative in the development of events in space at a certain time moment during a measurement. Initially and without any doubts, they propose to actually consider a quantum phenomenon in the frame of classical physics. Thus, the unconscious settings are used as early as the planning of the experiment and further determine the seeming paradoxicality of the result. The experimenters have no doubt that a cause-and-effect relation, which is determined by the sequence of events arranged in time one after one in an appropriate order, must exist in reality. This conviction relies on the fact that the presence of a spatially similar interval between events excludes such relation since there is no way to transmit the *information* at a velocity exceeding that of light. The meaning of this consists in that the events separated on the time scale by Δt interval and located in space at a distance of $\Delta x > c \times \Delta t$ cannot be causally related. This particular point reflects the unconscious setting implying that the actual reality is identical to our object-based model, where each object is defined in terms of its boundaries, is self-contained, and is independent.[10] Thus, it turns out that the model of reality as an integral system retaining its consistency at any level and in all cases, even when it decays (is measured) is simply out of the question. We stubbornly, to say the least, continue to cling to simple and illustrative classical representations of the world despite that physics has provided us with the experimental evidences that demonstrate our fallacies.

Then, they considered in the delayed-choice experiments the variant when a photon having passed the entrance semitransparent mirror of an interferometer encounters the situation when the presence or absence of a similar mirror at the exit is still not determined. In this case, the behavior of a photon as a wave (passing via both "arms" of interferometer) or a particle (passing either way) is still not determined by the observer and, thus, would puzzle Mother Nature.

"After the first half-silvered mirror (beam splitter) on the left, there are two possible paths, indicated by '2a' and '2b'. Depending on the choice made by the observer, different properties of the photon can be demonstrated. If the observer chooses to reveal its particle nature, he should not insert the second half-silvered mirror (½S), as shown at the bottom left in Fig. (**5.1**). With perfect mirrors (A and B) and 100% detection efficiency, both detectors will fire with equal probabilities but only one will fire for every individual photon and that event will be completely random. As Wheeler pointed out, "…one counter goes off, or the other. Thus the photon has traveled only *one* route" [68].

The experiment is based on the assumption that the behavior of a photon and its way in space—will be determined depending on the configuration of the used device, which is chosen at a certain time moment. Moreover, the elements themselves are thought of as objects rather than a whole system and, consequently, interference is not the feature of the system but rather the feature of its individual elements. Thus, the "quantumness" is ascribed to photon, which now possesses a wave–particle duality. This is completely confirmed by the below conclusion:[11]

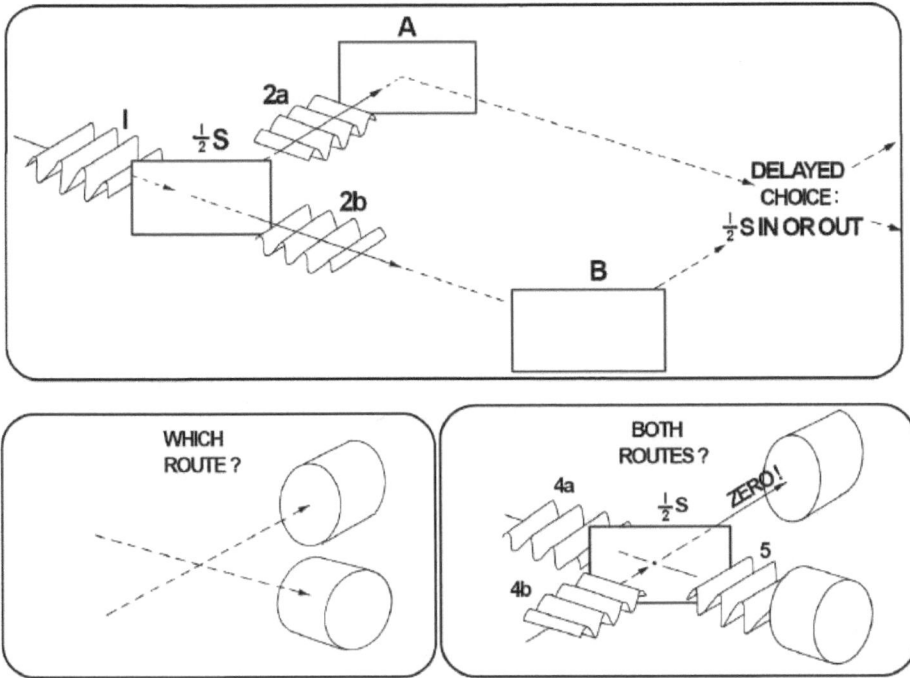

Fig. (5.1). Wheeler's delayed-choice gedanken experiment with a single-photon wave packet in a Mach–Zehnder interferometer. Top: The second half-silvered mirror (½S) of the interferometer can be inserted or removed at will. Bottom left: When ½S is removed, the detectors allow one to determine through which path the photon propagated. Which detector fires for an individual photon is absolutely random. Bottom right: When ½ S is inserted, the detection probabilities of the two detectors depend on the length difference between the two arms. The figure is taken from the study [68].

"Therefore, whether the experimenter's choice is in the time-like future of the emission event or space-like separated there from depends on the size of the interferometer and the amount of the time between the choice event and the photon arrival at the second beam splitter. Depending on the specific parameters, Wheeler's delayed-choice can thus be thought of as the time-like future of, or space-like separated from, the photon emission" [3].

"Therefore, important requirements for an ideal delayed-choice wave-particle duality experiment are (1) a free or random choice of measurement with space-like separation between the choice and entry of quantum system into the interferometer, and (2) using single-particle quantum states" [3].

The entire logic of the above reasoning is based on the dogma that time is real[12] and, consequently, the events must be related in a cause-and-effect manner. Correspondingly, any choice is followed by a response of the system that changes its behavior. This is an illustrative example of how an attempt to introduce a boundary in one way or another into a single entity (a systems-based representation) leads to a conflict with our object-based settings (as in the in vivo and in vitro experiments), the revision of which is most likely very difficult.

The further development of the ideas aiming to test the wave-particle duality (actually, the appropriateness of an object-based space-time description of quantum systems) was associated with how the spatial position of radiation source was recorded. It is clear that if we know the travel path of a photon, the interference is absent since this excludes the superposition of states. It is possible to design the experiment so that an alternative choice still exists after the photon is recorded, namely, the "travel path" (or the spatial position of the emitting atom) is either recorded or erased (the information about the atom that emitted the photon is removed). We expect that the measurement result should not change depending on the choice made a posteriori. Thus, if we in the first experiments as if "forced" a quantum system to guess its future state,[13] now we as if "force" it to revise its past. Note that the cause underlying this revision is the fact that we deliberately fix the final result of an experiment but **make before the choice** on whether to save or erase the information about the "travel path". We constantly consider the situation in terms of an object-based model by using a space-time description of a quantum system. However, the experimental results demonstrate that we fail "to cheat" reality [69–71]. Presumably, we cheat ourselves by imposing our own inadequate settings on Mother Nature.

Another set of experiments was focused on the purely quantum property, such as entanglement. There, the delayed-choice algorithms with entanglement swapping of two quantum systems were used. Of interest also is the fact that the overall situation is again described as usually in an object-based form. However, the nontrivial result itself requires a separate comment. Thus, the authors [3] write (Fig. (**5.2**) is copied from the same paper):

"When two systems are in an entangled quantum state, the correlations of the joint system are well defined but not the properties of individual systems [25, 72]. Peres raises the question of whether it is possible to produce entanglement between two systems *even after* they have been registered by detectors" [73] but the result remains unknown to us.

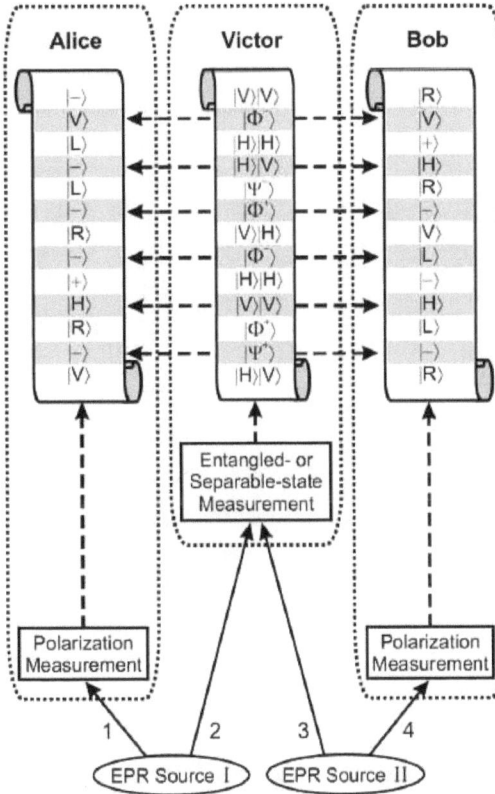

Fig. (5.2). The concept of delayed-choice entanglement swapping. Two entangled pairs of photons 1&2 and 3&4 are produced, *e.g.*, in the joint state $|\psi^-\rangle_{12}|\psi^-\rangle_{34}$ from the EPR sources I and II, respectively. Alice and Bob perform polarization measurements on photons 1 and 4 in any of the three mutually unbiased bases and record the outcomes. Victor has the freedom of either performing an entangled- or separable-state measurement on photons 2 and 3. If Victor decides to perform a separable-state measurement in the four-dimensional two-particle basis $|H\rangle_2|V\rangle_3, |V\rangle_2|H\rangle_3, |V\rangle_2|V\rangle_3$, then the outcome is random and one of these four product states. Photons 1 and 4 are projected into the corresponding product state and remain separable. On the other hand, if Victor chooses to perform an entangled-state measurement on photons 2 and 3 in the Bell-state basis s $|\psi^+\rangle_{23}, |\psi^-\rangle_{23}, |\Phi^+\rangle_{23}, |\Phi^-\rangle_{23}$, then the random result is one of the four Bell states. Consequently, photons 1 and 4 are also projected into the corresponding Bell state. Therefore, entanglement is swapped from pairs 1&2 and 3&4 to pairs 2&3 and 1&4. Figure adapted from Ref. [74].

When the question is put in this way, it is necessary to pay attention to the following fine point: Does a wave function collapse of the entangled pair (for example, as $\psi_0 = \frac{1}{\sqrt{2}}\left(|00\rangle + |11\rangle\right)$) after the first measurement means a complete "objectification" of the quantum system? There is no way to determine the very moment of the collapse in the model of reality, namely, whether it takes place during the measurement or during the realization of the measurement results by the observer. However, the fact that entanglement can arise between the systems that did not interact in this experiment after the first measurement suggests that the "decay" of the wave function of entangled pair ψ_0 does not mean that the "quantum" properties of the system completely disappear. This suggests that the notions "past–present" are inapplicable to reality. The axis of time does not emerge at the moment of measurement and the very event of "erasure" of the past is related only to our object-based model describing the result of measurement, while the realization of this is a mere fixation of the fact in the mental map with the corresponding remodeling of the space-time translation channel. [14]

Then, the authors describe the setup of the experiment by Peres for two pairs of entangled photons (1–2) and (3–4). [15]

"Peres suggested an addition to the entanglement-swapping protocol, thereby combining it with Wheeler's delayed-choice paradigm. He proposed that the correlations of photons 1 and 4 can be defined even after they have been detected via a later projection of photons 2 and 3 into an entangled state."

"According to Victor's choice and his results, Alice and Bob can sort their already recorded data into subsets and can verify that each subset behaves as if it consisted of either entangled or separable pairs of distant photons, which have neither communicated nor interacted in the past. Such an experiment leads to the seemingly paradoxical situation, that "entanglement is produced a posteriori, after the entangled particles have been measured and may even no longer exist" [73]."

"Since the property of whether the quantum state of photons 1 and 4 is separable or entangled, can be freely decided by Victor's choice of applying a separable-state or Bell-state measurement on photons 2 and 3 after photons 1 and 4 have been already measured, the delayed-choice wave–particle duality of a single particle is brought to an entanglement–separability duality of two particles" [3].

A positive result of such experiments completely contradicts the concept of reality as a chain of events ordered in time and separated in space. Moreover, a special focus on space-time diagrams in the relevant papers [75–77], first, demonstrates that there is no alternative to the object-based model when considering these experiments and, second, that it is completely irrelevant in this case.

Later in this review [3], the authors tell, "Five experimental runs with different space-time configurations were implemented. … The authors conclude: "The predictions of quantum mechanics are confirmed even with the choice of the final configuration being made randomly during the course of the 'elementary quantum phenomenon' " [78].

Finally, the authors [3] summarize the research into the manipulations with quantum states that have utilized the methods, such as delayed choice, quantum eraser, and entanglement of states and conclude: "Wheeler's delayed-choice experiments challenge a realistic explanation of the wave-particle duality. In such an explanation, every photon is assumed to behave either definitely as a wave (traveling both paths in an interferometer) or definitely as a particle (traveling only one of the paths), by adapting a priori to the experimental situation. Especially when the choice of whether or not to insert the second beam splitter into an interferometer is made space-like separated from the photon's entry into the interferometer, this picture becomes untenable."

"In delayed-choice experiments with two entangled quantum systems such as the delayed-choice quantum eraser, one can choose that one system exhibits wave or particle behavior by choosing different measurements for the other one. These choices and measurements can be made even after the former system has already been detected" [3].

In conclusion, the authors, while interpreting the experimental results explain the meaning of the term "quantum state": "…the state is a probability list for all possible measurement outcomes and **not a real physical object.** The relative temporal order of measurement events is not relevant, and no physical interactions or signals, let *al*one into the past, are necessary to explain the experimental results" [3].

The absence of understanding of the quantum mechanics is determined by the fact that our entire life goes within an object-based model. Actually, measurement is the creation of the conditions allowing an object-based model to be used for describing the part of reality that we distinguished as a quantum system. We have no other possibility to study its properties except for the decomposition of this system by reducing the wave properties in order to fit in with an object-based model. We perform space-time localization by unfolding a chain of events into a certain process. In fact, any fragment of the process at both microlevel and macrolevel is reversible. However, the development of events in a "reality–observer" system, in general, cannot be time-reversible since time is an attribute of the model of reality that possesses memory, that is, the observer.

It is necessary to note that Blokhintsev, as early as 1966, most correctly described the sense of the measuring process only without reference to the modeling of reality: "A certain macroscopic event always takes place at the output of any device, be it a turn of a counter arm, formation of drops in a cloud chamber, blackening of grains in emulsion, or the like. The quantum theory identifies this macroscopicity of a measuring device with the notion of its classicality; in other words, a measuring device must have the design allowing its classical features only to be eventually used for its operation…" [53].

Summing up, we have to state that the overall logic of experimenters when they do their experiments relies on an object-based space-time model. Moreover, they come to a completely correct inference that a quantum system cannot be regarded as a "physical object" and that space and time play no part in the attempt to explain the obtained results. Thus, we have to take only one small step and recognize the fact that we exist in a virtual environment generated by our unconscious and, therefore, we need to "divorce" our *representations* of reality and the reality itself. Poincare was perfectly right in indicating the mental character of the concepts of space and time. If each individual as a *specific observer's reference frame* mentally simulates the surrounding world as some objects distributed in space-time, there is no question on their unmistakable *relativity* for all remaining observer's reference frames. However, the Lorentz transformations, in this case, reflect a subjective perception similar to the perspective distortions rather than belonging to reality per se.

Gisin was absolutely right when he wrote in his monograph [65] that quantum nonlocality is a key point in physics. Nonlocality puts an end to our beliefs that space-time representations are real. The world does not depend on the methods we use to model it, yet the methods determine the limitations of its understanding. Gisin refers to EPR pairs as "entangled objects," and this is not a slip in speech. Our consciousness has not shaped any model of a quantum system and thus, our language lacks the corresponding notions. Presumably, if we are able to change our views on reality, community development may follow quite a different way, as exemplified by the disappeared civilizations of the East.

CONCLUSION

Есть усталости примета – настоящего не жаль,
И, как цифры под портретом, коротка о том печаль.

> *No mourn for the present is a sign of being tired.*
> *This sorrow is short as dates beneath a portrait.*
> *Midnight thoughts*

It looks as if the time we live in is just amazing: humankind during the last century implemented all fantasies described in science fiction novels by Jules Verne (*Twenty Thousand Leagues under the Seas*), Aleksey Tolstoy (*The*

Hyperboloid of Engineer Garin), Herbert Wells, and many others. We were taken up with the possibility to predict the future and imagine the trends in the development of technology and society. Among other things, we were absolutely sure that man had reached the summit not only of human development, but also the apex and end of the overall evolution. Most likely, the technological advance of our civilization made us believe that the human perception of reality is infallible and our ability to "customize" the reality to ourselves is boundless. However, the advent of quantum physics brought about some problems that bothered a few scientists who felt that the familiar views towards the surrounding world were coming to an end. It seemed as if we followed the wrong way, bringing forth manifold searches of the frontier area of physics, namely, an "observer–reality" system. This is especially evident in the remarkable books by Penrose [7, 37, 56, 79], which discuss these issues in a most open and consistent manner. How has it come that man emerged to be unable to perceive the own evolutionary biological legacy? How have we come to the attempts to fit reality to the Procrustean bed of our unconscious mindset? And why is it so difficult to revise our views despite that the experiments mercilessly lead us to the only conclusion without any chance to avoid it?

It is now quite evident that the more sophisticated the experiments in quantum physics, the more convincing will be the proof that our view of the world in the form of objects existing in space-time fails to fit reality. Correspondingly, any scientific modeling requires a very accurate use of these notions in order to avoid both the paradoxes and the pointless efforts to describe the world with "improper" methods in terms of only one model. Different metaphors must not be "blended" in an *adequate* description of reality; otherwise, we will get what we already have.

NOTES

[1] Definition of the object is absent in the encyclopedia of physics, while a physical understanding is important for us now; however, the term "matter" is explained as follows: "Matter is a fundamental scientific concept associated with any objects existing in nature, which we can estimate thanks to our perception. Physics describes matter as something existing in space and in time (in spacetime), which is the concept originating since Newton (space contains things and time contains events), or as something that self-specifies the properties of space and time, which originates from Leibnitz and is further expressed in Einstein's Theory of General Relativity."

[2] The category "Reality" is also absent in the physical literature, for example, such as *The Encyclopedia of Physics,* A.B. Prokhorov, ed., Moscow: Sovetskaya Entsiklopediya, 1988–1999.

[3] The spins of electrons in a common Euclidean space are anti-collinear (they are regarded as oriented along and against the magnetic field) versus a Hilbertian space, where they are orthogonal. This should be kept in mind, although the logic of reasoning does not change.

[4] Similar experiments are analyzed in detail in the monograph *Quantum Chance, Nonlocality, Teleportation and Other Quantum Marvels* by N. Gisin [65].

[5] Generally speaking, this contradicts the quantum mechanical description of this case since the state of the quantum system that we refer to as an "entangled pair" according to an object-based representation is put down by a single wave function. On the other hand, according to our assumption, each particle can be described by its own wave function. Thus, this variant in a sense is redundant.

[6] The question on whether the spin of the second particle evolves after the first particle is measured in this case is of minor importance since it raises the problem of information transmission rate, i.e., the problem of locality, which will be separately considered for the second variant.

[7] In its nature, the process of measurement is the translation of quantum system characteristics into an object-based model of reality, as is described in [6].

[8] Ten interpretations of quantum mechanics exist now [6]; however, this does not make it clearer.

[9] In general, this problem can be more properly formulated in the following manner: which particular conditions determine the applicability of an object-based or a systems-based model to the observed Reality.

[10] In essence, quantum mechanics implements a Platonic model of the Reality, in which "ideal" objects are not subject to any changes. They only can disappear or emerge (jumpwise) *in our perception* (in the consciousness, as in Plato's cave) under certain *observation* conditions as self-sufficient entities. However, the Reality itself is always unchangeable. As for us, we (except for Einstein) usually do not ask the question whether the Moon really exists when we close our eyes and do not look at it.

[11] It is assumed here, as usual, that the observer is not part of Reality.

[12] The only form existence, a spacetime one, is a priori imposed on Reality.

[13] Although it would be more logical to speak about the state of the subject.

[14] Then the presence of the Subject in the form of *Homo sapience* is not a necessary condition for existence of the Universe.

[15] When discussing such experiments with entangled pairs, characters named Alice, Bob, and Victor are introduced to entanglement swapping experiments to avoid impersonal phrases; these characters manipulate the states of quantum systems, for example, select a particular measurement basis.

CHAPTER 6

Coupling of the Models of Quantum and Classical Realities

Последний миг продлится вечность и, постигая бесконечность.
В мир, что творил, в его пределы уйдет душа из тела.

> *The final moment will be of an eternal length*
> *And soul will leave the body for the world created*
> *with comprehension of infinity.*
> *Midnight thoughts*

Abstract: The work by P. Florensky on the extension of geometry from the real region to a complex one under the metaphor of two non-overlapping worlds, real and imaginary, is successively analyzed. Assuming that consciousness functions in terms of a spacetime object-based concept, which actually corresponds to "classical reality", it is possible to hypothesize that the unconscious is related to the space of states and corresponds to a quantum description from a physical standpoint.

Keywords: Classical and quantum models of reality, Space of states, Special theory of relativity.

INTRODUCTION

We have earlier analyzed the capacities of a systems-based approach to the problem of coupling of consciousness and reality. This problem emerged in the most relevant way in the first half of the last century owing to the advent of quantum mechanics. The point is that the new theory radically differed from the classical one in that it gives only a probabilistic description of potential measurement (or observation) results for the quantum "objects", which already at that time was rather rejected by several physicists (according to Einstein, "God does not play dice"). This destroyed the regular object-based spatiotemporal worldview since a quantum object could simultaneously "reside" in different sites and "lose" its properties. Let us consider this in more detail.

Sergey P. Suprun, Anatoly P. Suprun & Victor F. Petrenko

In quantum theory, wave function Ψ, describing the state of a system, changes in a continuous manner; note that the full system retains its pure state, while the state of each subsystem becomes mixed. For example, the system comprising three subsystems—*X,* a studied object; *Y,* a measuring object; and *Z,* an "observer" (a physical body with the receptor machinery, peripheral nervous system, brain, and so on)—is described by the function $\Psi(x, y, z)$, $\Psi(x,y,z) = \sum C_k u_k(x) v_k(y) w_k(z)$, where C_k is the coefficient that reflects the probabilities for implementation of different states during measurement.

As is evident, the function $\Psi(x, y, z)$ is a superposition of the mixed states and does not give unambiguous information about object *X.* However, a measurement brings about a reduction of the wave function—a kind of "collapse" to a particular state, $u_k(x) v_k(y) w_k(z)$, which is not described by the theory at all. Since, as we see, the interaction with the nervous system of observer *Z* has nothing to do with this reduction, the measurement process actually ends at the moment when the situation is represented in observer's consciousness; correspondingly, it is reasonable to associate the factor that causes this reduction with consciousness. This created a lot of confusion among physicists; in particular, von Neumann in his *Mathematical Foundations of Quantum Mechanics* directly stated that this violated the principle of parallelism, which earlier allowed the physical processes to be considered separately from the observer's consciousness [2]. In this regard, Pauli believed that only the solution of a psychophysical problem could clarify this situation and put high hopes on joint work with Jung as a psychologist on the elaboration of the concept of *collective* unconscious [80] since this problem was either unsolvable at the level of *individual* consciousness or was reduced to pure solipsism.[1] However, this approach forces us to admit the *interaction* between consciousness (or unconscious) and physical reality and thus attribute it to the same reality, which would automatically exclude it from reduction factors. Perhaps, that is why Bohr asserted that *it is not reality* that quantum mechanics describes but only its potentiality; as for the reality itself, it appears *only at the moment of measurement* [4]. In this case, we obtain objective "non-reality" with all ensuing paradoxes (including the question of what is the entity with which an observer interacts in the situation of quantum superposition).

As we see, the attempts to solve the psychophysical problem and fit two realities of different types failed because of the natural opposition inherent in their very definitions. This contradiction disappears if consciousness and physical reality are considered within one system (since both must somehow be incorporated into the integrated Reality); moreover, the postulated parallel reality "beyond sensations" immediately appears to be redundant.

Note an important point that quantum mechanics[2] gives the most complete description of physical reality, while classical physics is in essence a mathematical formalization of what is represented (or what can be, at least in part, represented) in our consciousness. It is possible to make the views of Pauli, Jung, and Schrödinger consistent by associating the quantum reality with the unconscious and the classical one with the consciousness. Indeed, the only signal source of information about reality is our sensations, whereas the physical world "beyond sensations" is nothing but a hypothesis bringing about manifold contradictions and paradoxes. Assuming that the consciousness and collective unconscious are subsystems of the integrated Reality and avoiding "multiplication of entities without necessity" by devising a parallel "transcendent" world, we have to admit that what emerges in our consciousness originates from what is not consciousness, *i.e.*, from the unconscious.

In fact, certain content of Reality is translated via the subsystem of collective unconscious into individual consciousness as one of the subsystems of reality. Note that the content may be presented in different forms both as a whole and as a process: for example, a gramophone record, carrying certain fixed content, and playing on a phonograph. Another example is the information contained in DNA and the vital processes implementing its content with the help of cellular machinery. In this case, a subject at the level of the unconscious passes into a certain new state, while the psychic (brain) mechanisms "sweep" it into the corresponding processes (sensations) accessible to our consciousness and form the corresponding picture of "reality".

Thus, an object-based space-time method for representing reality emerges only in the subsystem of our individual consciousness, where the overall content is implemented as processes and motion is fundamentally "indestructible" since this is the only form for presenting the content of the subject's state in consciousness. Naturally, the systems-based evolution of our consciousness has led to the development of the psychic ("brain") functions that enabled the implementation of an "object-based folding" (as the most efficient) of the received information and its representation in the "physical space" constructed for this purpose.

This brings about the question of whether the space constructed in the consciousness by the mind for an object-based representation of reality can be regarded as physical. The paradoxes of quantum mechanics clearly suggest that it is inadequate to ascribe an object-based form to the Reality. As J. Robert Oppenheimer put it (cited according to [81]), "If we ask … whether the position of the electron remains the same, we must say "No"; if we ask whether the electron's position changes with time, we must say "No"; if we ask whether the electron is at rest, we must say "No"; if we ask whether it is in motion, we must say "No." Here, it is high time to question whether the electron actually exists as an object. However, assuming that the states of reality (naturally, confined to our "observer's reference frame") are directly presented to our unconscious and are translated to our consciousness in a "convenient" object-based spacetime form, many paradoxes disappear [6].

It is possible to refer to the works of Poincare, who proves that the dimension and metric of spacetime are arbitrary (conventional) since the very space is only the form of perception of reality in the consciousness created by the mind [17]. Presumably, different forms of reality exist, as Poincare postulated, and they have their merits and flaws relative to the problems that we solve. Poincare noted that different groups of transformations are referable either to "external" space or to own "internal" changes. In particular, when observing all transformations within a field of vision, we distinguish individual stable constellations, *i.e.*, objects. It would be more beneficial to ascribe some changes in these objects to space, for example, to regard linear fractional transformations as the changes in perspective associated with the observer's position rather than with a transformation of the object itself or to relate a decrease in the object's size depending on the distance between this object and the observer to its own movement (with a stationary observer) in the third dimension. All this is a matter of taste although the pattern of three-dimensional perception is deeply rooted in our mentality. However, such transformation is theoretically applicable not only to space, but also to time. What can be the result?

We experience our identity particularly because time does not affect our "self". Regardless of the fact that our memory, abilities, skills, and the overall "external world" continuously change, we experience our identity at any time interval starting from birth (more precisely with the realization of selfhood). Our ego is as if "bracketed out" of the physical space and time. Theoretically, we could turn this representation "outside in". The result of such an experiment would be a destructed identity of the own ego. The world would be beyond space and time and perceived as some changes in ourselves. In a sense, we ourselves would become this world. Many barriers separating us from reality would disappear. Any intricate cognitive mechanisms for navigating the world would be unnecessary being replaced by new intuitions. However, such a transformation would demand doing away with our own identity and would require serious efforts aimed at overcoming the attitudes that have been accumulated over our whole life.

Interestingly, the creative acts are also associated with a temporal *instantaneous* restructuring of the worldview and are accompanied by the effects of losing the self; that is why they are frequently regarded as the so-called altered state of consciousness (ASC).[3] In such a creative act, we feel as if we just suddenly "saw" (or more precisely, experienced) the answer even without the attempt to solve the problem in a cognitive manner.

The first scientists to speak about insight as early as the 17th century were Descartes, Spinoza, and Leibniz. They paid attention to the fact that there were some truths that mind discovered using a direct intellectual vision or intuition rather than based on logical reasoning. This is actually the meaning of *intuitio* in Latin. It is possible at this stage to see the solution immediately and the thought can take on a finished form. In his *Rules for the Direction of the Mind,* Descartes asserts "…we ought to investigate what we can clearly and evidently intuit… for knowledge can be attained in no other way." Einstein also wrote that the jump to a new idea is intuitive: "For me it is not dubious that our thinking goes on for the most part without the use of signs (words) and beyond that to a considerable degree unconsciously. For how, otherwise, should it happen that sometimes we "wonder" quite spontaneously about some experience? This "wondering" seems to occur when an experience comes into conflict with a world of concepts which is already sufficiently fixed in us." As for Gauss, he wrote, "I have had my results for a long time: but I do not yet know how I am to arrive at them." Another quote concerning a theorem that Gauss tackled for many years: "Finally, two days ago, I succeeded—not on account of my hard efforts, but by the grace of the Lord. Like a sudden flash of lightning, the riddle was solved. I am unable to say what was the conducting thread that connected what I previously knew with what made my success possible" [82].

Evidently, the attempt to fully transfer the own "inner" experience of an altered state to others will be unavoidably transformed into a certain "external" object-based form accessible to consciousness with the inevitable loss of the quality and the information available before.

In the first volume, we demonstrate that the initial configuration spaces of our sensations and experiences in the psychosemantic research into consciousness meet the Minkowski metric, which corresponds to the spatiotemporal metric of Einstein's STR. This is associated with the threshold of sensations and natural laws of perception, in particular, their "relative" nature. Thus, we do not need any physical justifications for the properties of spacetime; the psychological ones are sufficient.

However, when considering rather a formalized subsystem of the spacetime object-based (*process-based,* which is classical from the standpoint of physics) representation of reality, which we related to consciousness, we also need to somehow formalize the boundary between consciousness and the unconscious to which we relate the space of a completely other type, *the space of states* (the field of quantum mechanics in physics, which is also quite formalized). This would make it possible to consider these two subsystems together and understand how they "interact" with one another.

Note that the quantum world is actually quite different "reality" where the imaginary unit (i) "marks" what is hidden from our perception. In particular, Schrödinger's equation $(i\partial_t + \partial_{xx})\psi(x,t) = 0$ contains imaginary time, which directly indicates that the quantum phenomena are unrepresentable in consciousness. In principle, an imaginary subspace can be transformed into a visually illustrative form: Pavel Florensky in his *Mnimosti (Imaginary Points) in Geometry* [83] attempted to expand two-dimensional images of geometry and construct imaginary spaces. He set himself the task "to find the place in space for imaginary images but taking away nothing from the real images, which already took their seats" [83]. In the concluding remarks to this book, Leonid Antipenko notes: "...he [Florensky] not just laid the foundations for "imaginary geometry", but also demonstrated how the language of geometry can be used to express the link between the universe that is given to us from outside, based on sensational experiences, and the universe which we judge from inside based on the supra sensational intuition" [83].

Florensky gives the following line of reasoning in his book:

The area of triangle *ABC* in analytical geometry is calculated via the determinant

$$\Delta_{ABC} = \begin{vmatrix} x_1 & y_1 & 1 \\ x_2 & y_2 & 1 \\ x_3 & y_3 & 1 \end{vmatrix},$$

where x_i, y_i are the coordinates of the vertices $A(x_1, y_1)$, $B(x_2, y_2)$, $C(x_3, y_3)$ (Fig. **6.1**).

However, the area of the same triangle *ABC* is

$$\Delta_{ACB} = \begin{vmatrix} x_1 & y_1 & 1 \\ x_3 & y_3 & 1 \\ x_2 & y_2 & 1 \end{vmatrix} = -\Delta_{ABC}.$$

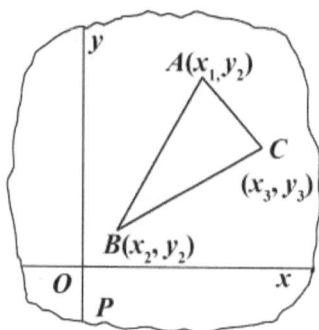

Fig. (6.1). Representation of a triangle in analytical geometry.

Thus, *the change in the order of rounding the angles of a triangle makes its area negative.* An analogous situation is when we look at the figure from below, *i.e.,* from under the plane. This means the existence of another, imaginary space (since the area became negative, all linear elements must become imaginary) on the other side of our real space. Florensky writes that "We can imagine the entire space as being dual, composed of the real and the corresponding imaginary Gaussian coordinate surfaces coinciding with the real ones; however, any transition from a real to an imaginary surface is possible only through a break in space and eversion of the body through itself" [83].

"Algebraic manipulations on complex numbers eventually lead to, and only to, complex numbers. Correspondingly, the coordinates of the points obtained from the equation of any plane curve can be real, purely imaginary, or complex and no other, that is, can be of only one of three following types: *a*, *ai*, or *a + bi*, where *a* and *b* are real numbers, positive or negative. Consequently, different combinations of the coordinates of these three types give the following possible nine types of the points on a plane as this is considered in analytical geometry and only these types; these nine types form six essentially different genera:

I.   *a, b* are real points;

II. $\left.\begin{array}{l} a, bi \\ ai, b \end{array}\right\}$ are semi-imaginary points

III.   *ai, bi* are imaginary points;

IV. $\left.\begin{array}{l} a, b+ci \\ a+di, b \end{array}\right\}$ are semi-complex points;

V.   *a + ci, b +ci* are complex points; and

VI. $\left.\begin{array}{l} a+di, bi \\ di, b+ci \end{array}\right\}$ are imaginary complex points.

The combinations joined with a brace belonging to the points that are similar to each other but have altered names of their coordinates. These are the six genera of the points of a plane (note that the points are "of a plane" rather than "on a plane")" [83].

The question arises on how the paired real and imaginary points can coincide. Florensky separated them by an infinitely small distance, which he identified with Leibniz's differential. However, Antipenko notes that "he [Florensky] at that time could not use in full the notion of infinitely small quantity in the relevant meaning, since it was elaborated several decades later by a new field of mathematics, namely, nonstandard or non-Archimedean analysis" [83]. It can be said that nonstandard analysis vindicates the inexact reasoning by Leibnitz, Bernoulli, Euler, and other classics. "Infinitesimals" and "infinites" are defined in the nonstandard analysis as static objects rather than sequences and functions tending to zero. These static objects are grounded by constructive analysis. This approach agrees well with both the intuition and real history of the origin of mathematical analysis.

Brief Information for Reference

The applications of nonstandard analysis cover a vast area from topology to the theory of differential equations, theory of measure, and probability theory. The use of nonstandard Gilbert's space in the construction of quantum mechanics looks promising.

As has been noted, the key point of nonstandard analysis is that it considers infinitesimals as a constant rather than as variables (*i.e.*, rather than the functions tending to zero). For example, let $\varepsilon > 0$ be one of such infinitesimals; then according to its definition, by adding ε to itself, we can get the numbers $\varepsilon + \varepsilon$, $\varepsilon + \varepsilon + \varepsilon$, $\varepsilon + \varepsilon + \varepsilon + \varepsilon$, and so on. If all numbers thus obtained are always less than unity, then ε will comply with the definition of infinitesimal. The existence of infinitesimals suggests the existence of infinitely large quantities, infinites. In particular, this is the reciprocal of ε, since it is always larger than any finite sum of unities, $1 < 1/\varepsilon$, $1 + 1 < 1/\varepsilon$, $1 + 1 + 1 < 1/\varepsilon$, ...

Thus, if ε is infinitesimal, then $1/\varepsilon$ is infinitely large in the sense that it is always larger than any number of 1, $1 + 1$, $1 + 1 + 1$, $1 + 1 + 1 + 1$, and so on. Thus, if we measure a segment with a length of $1/\varepsilon$ using a length standard (*i.e.*, successively marking the segments of unit length), we can never complete the process of measurement.

All this suggests that the existence of infinitely small quantities contradicts the so-called Archimedean property (measurement axiom), stating that for any segments

A and B, it is possible to mark on a line the smaller segment (A) as many times as to finally obtain a segment longer than the large segment (B).

The infinitesimals and infinites thus defined are referred to as *nonstandard* versus the common standard real numbers. According to the transference principle, all laws of standard algebraic manipulations (addition, subtraction, multiplication, division, *etc.*) are applicable to them with some reservations. The field of real numbers expanded at the expense of new numbers is referred to as the field of hyperreal numbers. The concept of a hyperreal number unites the standard and nonstandard numbers; note that the hyperreal numbers that are not infinitely large are referred to as finite. Both standard and nonstandard numbers belong to finite ones. Taking into account that zero meets the definition of an infinitesimal, each finite hyperreal number a can be expressed as a sum $b + \varepsilon$, where b is a standard number and ε, is infinitesimal. Number b is a standard part of hyperreal number a, which is put down as $b = \mathrm{st}(a)$.

The problem of existence of nonstandard numbers is solved in mathematics by the discovery of nonstandard models of arithmetic with hypernatural numbers. The difference between the standard and nonstandard axiomatics of mathematical analysis is determined by their relation to the Archimedean property.

Two hyperreal numbers are regarded as infinitely close if their difference is infinitely small. As follows from the properties of infinitesimals, the relation of infinite proximity is the equivalence relation. In particular, this means that the relation of infinite proximity is reflexive (each x is infinitely close to itself), symmetric (if x is infinitely close to y, than y is infinitely close to x), and transitive (if x is infinitely close to y and y is infinitively close to z, than x is infinitely close to z). As is known, each equivalence relation partitions the set where it is defined into pairwise disjoint classes so that any two elements belonging to the same class are equivalent and any two elements belonging to different classes are nonequivalent. In our case, the relation of proximity partitions the axis of hyperreal numbers into such disjoint classes that the elements belonging to each class are infinitely close to each other, whereas the elements from different classes are not. The classes containing standard real numbers are referred to as monads.

Considering the elliptic geometry of STR, Florensky writes: "However, these general geometric considerations have recently got an unexpected concrete interpretation in the principle of relativity, and the universal space from the standpoint of modern physics must be conceived exactly as an elliptic space and is recognized as finite as well as time, being finite and closed into itself" [83].

"The characteristics of the bodies in a moving system observed from a stationary one depend on the basic variable $\beta = \sqrt{1 - v^2/C^2}$, where v is the velocity of the system and c, velocity of light. Until v is smaller than c, β is real and all characteristics remain immanent to a terrestrial experiment; at v equal to c, $\beta = 0$; and at v larger than c, β becomes imaginary. A twofold qualitative jump of the corresponding characteristics takes place in the two last cases. In particular, in the moving system, the length of the bodies reduces at a ratio of β: 1 in the direction of movement; time, at a ratio of 1: β; mass, at a ratio of 1: β; and so on" [83].

"This is at the limit, at $\beta = 0$. However, beyond the limit, at $v > c$, time flows in the opposite sense so that the consequence precedes the cause. In other words, the acting causality here is replaced ... by the final causality, teleology—and the kingdom of goals stretches beyond the boundary of limit velocities. In this area, both the body length and mass become imaginary. However, keeping in mind the interpretation of imaginaries proposed here, we visualize how a body dwindling to zero falls through the surface—the carrier of the corresponding coordinate—and everting through itself acquires imaginary characteristics. Figuratively speaking, and in the case of a specific comprehension of space, not figuratively, we can say that space is broken at the velocities exceeding that of light, similar to the air that is broken when bodies move faster than the velocity of sound; this brings qualitatively new conditions for the existence of the space that are characterized by imaginary parameters. But similar to the fall of a geometric figure through the plane, which does not mean its annihilation but rather its transition to another side of the surface and, consequently, accessibility to the beings inhabiting the other side of the surface, an imaginary nature of the body's parameters must be understood not as a sign of irreality of this body but rather as an evidence that it passes to another reality. The region of imaginaries is real, perceivable, and in the language of Dante is called the Empyrean. We may imagine the entire space as being dual, composed of the real and the corresponding imaginary Gaussian coordinate surfaces coinciding with the real ones; however, any transition from a real to an imaginary surface is possible only through a break in space and eversion of the body through itself. Meanwhile, we envisage the only way to this process as the increase in velocities, perhaps, the velocities of some parts of the body, over the limit velocity of c; but there is no evidence that some other ways are infeasible" [83].

Antipenko in his comments writes, «Mathematics since ancient time deals with the studies of metric and topological characteristics of the spatiotemporal manifold. However, all traditional studies failed to address the problem posed by Florensky, namely, the question on the existence of the insideness, "underside" of spacetime. Presumably, this is determined by the fact that the assumption of the dual nature of spatiotemporal relations inevitably bumps up the very same antinomies that lacked any logical vindication before the works by Nikolai Vasil'ev.[4] Now the fear of contradictions as antinomies is definitely overcome and the very mathematical logic guides the search for unusual, intangible, and superempirical in the spacetime universum. The first step in this way is the discovery of polarization of the spatiotemporal relations into external and internal. The last step is the completion in the construction of Florensky's geometric model" [83].

In order to understand what underlies the polarization of spacetime relations into external and internal, it is necessary, first and foremost, to focus on the dual representation of relations in the language of logic, in extensional and intensional aspects. The relations and properties in mathematical logic are united under the (common) name of predicates. Property corresponds to a monadic, while n-place relation when $n > 1$ corresponds to a multiplace predicate. Both extensional and intensional aspects of relations are defined by the analogy to extensional and intensional of properties. The extensional of a property is represented as a set of objects (entities) possessing the corresponding property. As for the intensional aspect, the property is an attribute inherent in an entity. In an analogous manner, a set of ordered n objects is referred to as the extensional of n-place relation, while the intensional of relation takes into account the attributes of objects defining the relation between the distinguished objects [83]. "The case in point for a two-place predicate is the relation between the percipient (subject) and perceived entity (object), studied in psychophysiology and epistemology. Its sides are united by the notion of secondary qualities of things, *i.e.*, the qualities that depend on both members of a binary relation, namely, on the object and the subject. Any common view on the nature of the links between external and internal relations has not been established so far.

It is known that logical positivists—Russell, Wittgenstein, and others—recognized only external relations. As for the external ones, they were expelled from both logic and philosophy. When analyzing this issue, Bo. Dahlbom writes that James Moore and Bertrand Russell used external relations as the main weapon against idealism. Certainly, the central were the relations between the cognizer and the cognized entity, of percipient and his sensations with the perceived object, and so on, where the internal relations underlie the intimate side of idealism, which was lost when they believed that these relations are external" [83].

"Florensky [...] approached the dilemma according to which the points constituting the spatiotemporal manifold behave differently: in one case, they are disjoint and any part of the space is separable from an adjacent part; in other case, this is not possible because, once having a contact, any two adjacent points immediately "stick together". In topology, this dilemma is related to the concepts of separable (Hausdorff) and inseparable topological spaces. This is associated with a very important problem and his *Mnimosti in Geometry* leads to its definite solution" [83].

Space X is referred to as a Hausdorff space or separable space if for any two distinct points x and y there exist nonoverlapping open sets X and Y containing points x and y, respectively. In other words, space X is separable by definition if for any arbitrary points x and y of this set there exist the neighborhoods that have no points in common.

"What is the way that leads to satisfaction of the sufficiency condition for a topological space to be separable? Judging from the master plan of *Mnimosti in Geometry,* this way is genially simple. Two classes of standard points are introduced—the real points and the imaginary points—that are indexed with real and imaginary numbers, respectively. The real and imaginary points are placed on the number axis so that each imaginary point goes after a real point and vice versa, *i.e.*, they alternate. Such fitting of imaginary points into the number axis ensures that the monads of any two standard points do not overlap" [83].

In the pseudo-Euclidean geometry of STR, the spacetime interval Δs acts as the distance between two events:

$$ds^2 = c^2\, dt^2 - dx^2 - dy^2 - dz^2$$

A mirror inversion in the imaginary geometry of Florensky is the transformation $ds^2 \rightarrow -ds^2$ (*i.e.*, the shift by the value of infinitesimal ε). Correspondingly,

$$(ds')^2 = c^2\,(dt-\varepsilon)^2 - (dx-\varepsilon)^2 - (dy-\varepsilon)^2 - (dz-\varepsilon)^2.$$

If $(ds')^2 = -ds^2$, than for dx, in particular, $-dx^2 = dx^2 - 2\varepsilon dx + \varepsilon^2$ or $dx_{1,2} = (1\pm i)\,\varepsilon/2 = \dfrac{\varepsilon\sqrt{2}}{2}(\cos\theta \pm i\sin\theta)$, where $\theta = \pi/4$. At $\theta = 0$ or $\pi/4$ the real and imaginary points are separated so that the space becomes topologically inseparable: the uniform points "stick together" and form indivisible medium of instantaneous effects that are independent of the distance between the objects. The same is for dy, dz, and dt.

In the case of standard manipulations on complex numbers, $\mathrm{Re}(dx) = 0$ at $\theta = \pi/2$. However, considering the manipulations on complex numbers in the non-Archimedean analysis, the radial vector of a complex number in the Cauchy plane may coincide with the abscissa or the ordinate only in the limits allowed by the value of infinitesimal ε. Correspondingly, the equation for dx at $\theta = (\pi/2 - \varepsilon)$ takes on the following form:

$$dx_{1,2} = \frac{\varepsilon\sqrt{2}}{2}\left[\cos\left(\frac{\pi}{2} - \varepsilon\right) \pm i\sin\left(\frac{\pi}{2} - \varepsilon\right)\right] \approx i\frac{\varepsilon\sqrt{2}}{2}.$$

Since the real component of complex variable dx is absent there, the distance between the two closest real points disappears (*i.e.*, becomes imaginary). Thus, the overall real manifold becomes single and indivisible, resembling a perfect elastic body, capable of transmitting acoustic perturbation with an infinite velocity.

In a mirror world, the radial vector of a complex number on the Cauchy plane must turn clockwise rather than counterclockwise starting from $\theta = \pi/4$. The points disjoin at $\theta = \varepsilon$. Consequently, the coordinate structure of four-dimensional spatiotemporal Minkowski manifold in an integrated form looks like

$$\begin{cases} \vec{r}_c = \vec{r} + i\vec{r}' \\ t_c = t + it' \end{cases}$$

For our real world, $\vec{r}' \equiv t' \equiv 0$, and for the mirror world, $\vec{r} \equiv t \equiv 0$; r and t here run from $+\infty$ to $-\infty$.

Consider as an example the transformation of Schrödinger's equation $\dfrac{\partial \psi}{\partial t} = \dfrac{i\hbar}{2m}\Delta\psi$ in the case of a mirror reflection in Florensky's geometry. In a one-dimensional case, $\dfrac{\partial \psi}{\partial t} = \dfrac{i\hbar}{2m}\dfrac{\partial^2 \psi}{\partial x^2}$. Making the replacement $\begin{cases} \partial t^2 \to -\partial t^2 & \Rightarrow & \partial t \to i\partial t \\ \partial x^2 \to -\partial x^2 & \Rightarrow & \partial x \to i\partial x \end{cases}$.

we get the diffusion equation (or heat conduction equation)

$$-i\frac{\partial \psi}{\partial t} = -\frac{i\hbar}{2m}\frac{\partial^2 \psi}{\partial x^2} \quad \Rightarrow \quad \frac{\partial \psi}{\partial t} = D\frac{\partial^2 \psi}{\partial x^2}.$$

Commonly, the diffusion equation appears from the empiric equation stating proportionality of the flow of matter (or heat energy) to the difference in concentrations (temperatures) of the regions separated by a thin layer of a substance with a given permeability, characterized by diffusion (or heat conduction) coefficient.

In a one-dimensional case, the fundamental solution of homogeneous equation with constant D (at the initial condition defined by the delta function

$\psi_f(x, 0) = \delta(x)$ and boundary condition $\psi_f(\infty, t) = 0$ is $\psi_f(x, t) = \sqrt{\dfrac{1}{4\pi Dt}} \exp\left(-\dfrac{x^2}{4d}t\right)$.

In this case, $\psi f(x, t)$ is interpretable as the density of the probability to find a particle that at the initial time moment was at the initial point at the position with coordinate x at time t. Analogously to Schrödinger's equation: "An important part of the Schrödinger equation for a single particle is the idea that the probability to find the particle at a position is given by the absolute square of the wave function. It is also characteristic of the quantum mechanics that probability is conserved in a local sense. When the probability of finding the electron somewhere decreases, while the probability of the electron being elsewhere increases (keeping the total probability unchanged), something must be going on in between. In other words, the electron has a continuity in the sense that if the probability decreases at one place and builds up at another place, there must be some kind of flow between. If you put a wall, for example, in the way, it will have an influence and the probabilities will not be the same. So the conservation of probability alone is not the complete statement of the conservation law, just as the conservation of energy alone is not as deep and important as the *local* conservation of energy. ... If energy is disappearing, there must be a flow of energy to correspond. In the same way, we would like to find a "current" of probability such that if there is any change in the probability density (the probability of being found in a unit volume), it can be considered as coming from an inflow or an outflow due to some current. This current would be a vector which could be interpreted this way—the x component would be the net probability per second and per unit area that a particle passes in the x-direction across a plane parallel to the yz plane. Passage toward $+x$ is considered a positive flow, and passage in the opposite direction, a negative flow" [85].

Note that the fundamental solution of the heat conduction equation inherits some peculiarities of the microworld of quantum reality: an instantaneous heat pulse gives an instantaneous increase in temperature at any point of the space independently of its position, *i.e.*, heat propagates with an infinite velocity, which is an evident nonsense in classical physics, being similar to the situation with the velocity of a diffusing substance [86] and is a typical event in quantum mechanics.

CONCLUSION

Thus, the approach of Pavel Florensky to the extension of geometry from the real to complex one gives good prospects for integrating the two subsystems of reality—consciousness and the unconscious, and the classical world and the quantum world.

NOTES

[1] The known paradox of Wigner's friend consists in that different observers perceiving a quantum system can in principle record different states; however, this does not occur.

[2] Which according to Pauli is completely unrepresentable in our consciousness [4].

[3] Although it is more correct to speak about an altered state of *subject*.

[4] For example, [84].

CHAPTER 7

Purposeful Systems, Evolution, and Subject-Based Aspect of Systems Science

"Imagination is the only weapon in the war against reality."
Lewis Carroll
Alice's Adventures in Wonderland

Abstract: In systems science, the category of goal is restored in its rights. However, the current studies are mainly confined to closed systems and are of an object-based character although the teleological principle, defining a system, is inapplicable to the object. The issues associated with the possibility of the subject to be included in systems analysis are discussed as well as the approaches to the research into evolutionary processes in open systems. In addition to this, the problems in a subject-based approach in psychosemantics and quantum physics are considered.

Keywords: Evolutionary processes, Mental map, Open systems, Semantic space, Subjective and objective approaches, Systems science, Teleological principle.

INTRODUCTION

Systems science relies on the view of the world as a hierarchical, ordered diversity. Historically, three periods are distinguishable in the development of the systems approach. The characteristics of the antic period were mainly **(i)** a holistic view on nature associated with indivisibility of the primary elements considered at that time, namely, earth, water, air, and fire; **(ii)** a speculative (theoretical) approach separated with experiment; and **(iii)** wide use of the category of goal. In particular, the falling of a stone was explained by its wish to return to its place [87].

The speculations of antic science were replaced in the 17th century by the modern natural-science approach, first and foremost, associated with the name of Newton. Characteristics of the Newtonian period, still continuing, with its physicalist methodology are **(i)** combination of experimental and theoretical (mathematical) approaches; **(ii)** reduction of the study of a whole entity to examination of its

Sergey P. Suprun, Anatoly P. Suprun & Victor F. Petrenko

parts; and **(iii)** "expelling" of the category of goal although some rudiments of goal functions were retained, being reduced to the principles of nature (uppercased in this situation). The principles of physics (of least action, entropy increment, and so on) and biology (natural selection and so on) actually matched the goals ("needs") of nature as a subject and were in fact the explanatory categories when answering the question on why a particular phenomenon goes in this particular way, or, according to Einstein, to understand empirical regularity as a logical necessity. This approach implements the optimality principle (also known as the extreme or variation principle), stating a minimum (or a maximum) of a certain variable ("functional" or "*objective* function"). In optics, this is Fermat's principle of least time; in mechanics, the principle of least action; and in thermodynamics, maximum entropy principle. This particular teleological component of the extreme-based principles caused the strongest rejection when Pierre Maupertuis on April 15, 1744 presented the memoir where he offered a new universal principle of mechanics—the principle of least action, stating that the true motion differs from all possible variants in that the value of action for it is minimal. This memoir stirred up a heated debate among the scientists of that time, falling beyond the scope of mechanics. The main matter of dispute was whether the events taking place in the world are causal or teleologically guided by certain supreme intelligence *via* "final reasons", *i.e.*, goals. Maupertuis himself emphasized and defended the teleological nature of his principle and directly asserted that "the economy of effort" in nature proves the existence of God. Leaving aside the discussion of divine issues, we still need to note that when applying teleological principles, we regard Nature as a subject that implements certain goals. This is rather close to Einstein's statement: "I believe in Spinoza's God, who reveals Himself in the lawful harmony of the world, not in a God who concerns Himself with the fate and doings of mankind" [88].

Advent and Development of Systems Science

The modern systems approach has developed from **(i)** the understanding of unfeasibility to reduce all properties of a complex system to its elementary components (emergence); **(ii)** unfeasibility of classical experimental approach (for example, for complex unique systems); and **(iii)** upholding of the rights of the category of goal. Structural and behavioral qualities of complex systems rather than their material and energy-related characteristics have emerged to be decisive; thus, the former have become the main object in systems science. This approach of Russian psychology is usually associated with the works of Pyotr Anokhin [89], Nikolai Bernshtein [48], Aleksandr Luria [90], and others.

The general systems theory was proposed by Ludwig von Bertalanffy in the 1930s [13]. His main idea was that the laws determining the function of systems objects are isomorphic. He also introduced the concept of open systems, *i.e.*, the systems that constantly exchange matter and energy with environment, and studied them.

Bertalanffy himself believed that the following scientific disciplines have (in part) the common goals or methods with the systems theory:

1. *Cybernetics,* which relies on the principle of feedback;
2. *Theory of information,* introducing the notion of information as a certain measurable quantity and developing the principles of information transmission;
3. *Games theory,* which uses the special mathematical apparatus to analyze a rational competition of two or more counteracting forces aimed at attaining the maximum gain and the minimum loss;
4. *Theory of decision making,* which analyzes the rational selection within human institutions;
5. *Topology, which includes the nonmetric fields,* such as network theory and graph theory;
6. *Factor analysis,* that is, the procedures of separating factors in multivariate phenomena in psychology and other scientific areas; and
7. *General systems theory in a narrow sense,* attempting to derive several notions characteristic of the organized whole (such as interaction, sum, mechanization, centralization, competition, and finality) from the general definitions of the notion of "system" and applying them to particular phenomena.

The paradigm of systems approach formed as early as the mid-last century based on the studies by Singer [91, 92], Rosenblueth and Wiener [93, 94], and Sommerhoff [95]. They were the first who clearly realized the benefits of regarding the mechanisms as teleological functional entities. As the concept has emerged, it is conceptually more useful to move from functional integrity to structural components rather than in the opposite direction. Moreover, Singer has shown that the very structure is a functional notion. Before, the researchers in their attempts to understand the whole relied on the analysis of its components; now they try to get the insight into components by decomposition of their knowledge about the whole. This research algorithm is referred to systems analysis [96, 97]. Bohr believed that the mechanistic and teleological standpoints had no contradiction but were rather complementary. Einstein when speaking about the imperative of the modern science notices that we want to know nothing but understand everything. Evidently, a teleological approach is most fruitful when studying biological and social systems as well as human behavior, although the works in these areas are believed to be yet rather eclectic [98]. Russell in his *History of Western Philosophy* asserts that only mechanistic (cause-and-effect) approach is in line with the modern scientific paradigm and scientific knowledge

[99]. This position actually excludes the studies of any purposeful systems from scientific consideration since they are related to a teleological explanation.

Scientific progress and the advancement in science are accompanied by grouping of the phenomena into ever-narrower classes and development of the disciplines focusing on the research into each of these classes. However, Mother Nature herself does not appear being partitioned into disciplines; this is no more than our method to order our own knowledge. Disciplines represent a system of knowledge rather determined by points of view and the methods used for studying them. Evidently, *the very partitioning of the natural phenomena into classes is in many respects predetermined by our needs, motivations, and goals.* In this sense, the science itself as a system should be necessarily considered in terms of a functional teleological or purposeful approach.[1] Any class of objects distinguished by science in accordance with a certain goal is *nonrandomly* united by a common function and thus *the objects within this class display **a certain similarity.*** A mental *procedure of selection of objects and formation of the corresponding class* utilizes a certain *nonrandom factor* that links the descriptors for the objects of this class by a certain correlation. Correspondingly, *the relations of the properties of objects within this class are nonrandom.* This particular point makes *redundant a semantic description of objects via* their basic properties, since some of them can be defined *via* other objects based on these specific relations characteristic of a particular class of objects (their similarity). The way to avoid redundancy in the description of objects is the transition from basic properties (and the corresponding semantic units) to orthogonal factors that unite the synonymic description components in the corresponding class of objects, thereby considerably reducing the dimensionality of semantic space. Since each factor determines a certain regression equation that links the properties by a certain relation within a class of objects, we automatically also get the *specific laws for relation of properties* in this class.

Omitting the evolutionary aspect in the formation of needs, teleological concepts can be made similarly objective, measurable, and appropriate for formalization similar to the other structural concepts of classical science. As for the creative processes—evolution and creativity—these processes alter the very axiomatics of a formalized system and thus require some other approaches for their analysis. However, recent studies in biology, psychology, and sociology demonstrate that the creative processes are associated with radical qualitative changes, are jumpwise, and take place rather infrequently. A change in social and economic systems as well as in social needs and motivations is a rare historical event that cardinally changes the overall mentality of a society, while biological speciation is a rather slow process even as compared with the history of civilization. Since the intervals when these systems are "stationary" are rather long, their formalization within these boundaries does not contradict the "mechanistic" views of the classical scientific approach. Evidently, the steady state intervals decrease when a system passes to a new level of evolutionary development (from physical through biological to mental), increasing the influence of evolutionary processes. Starting from biology, the evolutionary processes represent an integral characteristic of these systems as distinguishing them from "inanimate" matter. However, researchers when describing evolutionary processes try to remain within the purely object-based mechanistic paradigm. Anokhin liked to quote a phrase by Mihajlo Mesarovich that teleology is a woman without whom no biologist can live but is ashamed to go out with her. Two types of teleology are distinguishable, subjective and objective. The creation of the former is attributed to Aristotle, who developed teleological notions to explain the behavior of things and animals. For example, a psychologist who deals with the notions, such as convictions, inclinations, instincts, and motives, to explain human behavior follows subjective teleology. As for objective teleology, these notions follow from what an individual does; they are objectively derived from what we can observe. All properties are derived based on the perception of the regularity in behavior under certain circumstances. These two approaches are consistent. Although human behavior is determined by "subjective" variables of a person, this behavior is interpreted in the frame of objective variables of the observed behavior since the subjective components are not directly perceived. The very existence of a social medium, which requires efficient communications and understanding of the motivations of its members, demonstrates that the interaction of two teleologies and the issues of "subjective" interpretation of "objective" variables are quite legitimate and amenable to objective scientific study and interpretation (at least within the same mentality).

Doubtlessly, the language of intentions, feelings, and sensations cannot be translated into the language of words and notions in full and without loss; however, this is not necessary for socially significant communications. Evidently, the integral behavior can be decomposed into individual components in different ways based on the relevant needs, motivations, and goals of a particular mentality. In fact, one of the goals of psychosemantics is the decomposition of a perceived phenomenon, communication, or behavior into individual gestalts and their reference to the subjective components of the perceiving individual himself\herself (or a particular mentality), *i.e.*, the problem of interpretation of cultural meaning and personal meaning. Indeed, communication as information transmission is divisible into smaller subsystems in Shannon's manner: encoding, transmission, data processing, storage, output, indexing, and so on. Note that Shannon [100] did not consider the person transmitting a communication in his model of communication although we must understand *why* and *what for* individuals communicate in a certain way when studying the communication processes. Thus, we come to the matter of choice and it has to become an integral part of any complete model of communication [97]. It is evident that the strategy of choosing and its variants are first and foremost determined by the relevant motives and states of an individual or a purposeful system. It is nonsense to interpret cultural meaning and personal meaning as invariant for all mentalities. Interpretation of a communication and further behavior of a system significantly depends on these factors. However, the researchers dealing with the systems, such as market and advertizing, as a rule regard individuals as the automata mechanically generating statistical data without taking into account the human factor and reduce the problem of choice to the rules of logical deduction [101].

The advent of cybernetics, informatics, general systems theory, studies of operations, and other related scientific directions increased the interest in teleological notions, such as the function and the goal. According to Miller, Galanter, and Pribram [102], these new trends influenced even the behavioral sciences, which traditionally had a mechanistic orientation. However, different disciplines now and then interpret these notions in a radically different manner, which does not enhance interdisciplinary studies into human behavior. Academician Vyacheslav Stepin, a recognized expert in the philosophy of science, explains this in the following way. *Classical science* relies on the principle of repeatability, objectivity, and independence of the time and place of observation. Neither the tool nor observer and his/her attitudes were essential. *Nonclassical science* has shown that the situation in many cases is more intricate. As it has emerged, we are unable to simultaneously measure the coordinate and momentum of a microparticle in quantum mechanics. Moreover, the very measurement procedure also changes the properties of the particle and of what will be measured. In addition, an inherent probability emerges as a result of numerous experiments. This is far from a stringent cause-effect relation as in classical mechanics. However, we ever more frequently encounter the next stage in cognition, namely, *postnonclassical science.* Here, the research object is equally essential as the used tools and the cognizing subject. A simple example is the notion of safety in the context of environmental conservation. It is evident that hares, wolves, and humans trying to conserve biodiversity interpret the answer to this question from different positions and have different worldviews [103].

In the view of Wiener and Rosenblueth [97], the behavior of a purposeful object should match certain specific features of the environment and "be goal-oriented and goal-directed". The main criterion of goal orientation is that a system or an object continues to pursue the same goal fitting its behavior to the changing environmental conditions. However, this purely cybernetic definition fails to cover all aspects of human behavior. Sommerhoff [104] has provided a good example of such discrepancy: "I approach the door of my house and enter. Here the approach towards the door might possibly be interpreted as an error-controlled movement in which visual and proprioceptive impulses provide the basis of an error computation which is then used in determining the corrective output. But it does not apply to my choice of this door as distinct from the other doors of the street … even if we just look at the movements involved in approaching the chosen door, all we can say strictly speaking on the observed facts of the case is that these movements are *error eliminating*—not that they are *error controlled*—in the strict sense of the term in which this implies a mechanism based on the initial computation and setting up of an error signal reflecting the magnitudes and direction of the discrepancy between the actual and desired state of the system. According to Bernshtein's concept [48], these "desired states" should be somehow represented in the engram and nervous system and are a kind of what Anokhin [89] referred to as the "acceptor of action".

Thus, manifold questions arise at the very beginning when defining what systems analysis is. Consider first the definition of a system as it is commonly given in various dictionaries: *System* (σύστημα in Greek, "the whole concept made of several parts or members") is a regularly interacting or interdependent group of items forming an integrated whole [105]. Note that any nonelementary object can be considered as a subsystem of the whole (to which the object in question belongs) by distinguishing separate components of this object and determining the interaction of these components that perform a certain function. When using this definition, any theory in principle gives a systems-based description because it first and foremost considers the relations between the objects, their interaction, and their patterns. Although the systems science emphasizes several properties, such as *integrity, hierarchy,* and *emergency,* they are frequently only announced, whereas the actual research is reduced to an object-based scheme. However, if we introduce the category of *goal, motivation,* or *need,* we should keep in mind that they are relevant to the *subject* rather than to the *objects;* thus, the systems analysis is necessarily *subject-based.* Only subject (*subjectus* in Latin, "lying beneath") as a "source of activity directed towards object" can form the background of systems science. The subject does not complement the object but rather opposes it (*objicio* in Latin, "to throw against"). The object-based approach decomposes reality with a subsequent analysis and synthesis of the separated units and relations by describing their *meaning* and gives the answer to the question of *how this works,* whereas the subject-based approach refers to the goals and motivations, reveals their *meaning*, and answers the question of *why* and what for something exists. In particular, the explanation of motion kinetics of an individual in the earlier considered example by Sommerhoff fails to answer the question of n *why* he selected this particular direction.

The attempt to mechanistically unite subject-based and object-based approaches within a single concept is most likely unfeasible. Interestingly, the semantics of the category "subject" corresponds to semantics of the category "substance" (**substantia** in Latin, "*something that forms the foundation*"; compared with **subjectus,** "lying beneath"). Russell writes that substance "...when taken seriously, is a concept impossible to free from difficulties. A substance is supposed to be the subject of properties, and to be something distinct from all its properties. But when we take away the properties, and try to imagine the substance by itself, we find that there is nothing left... "Substance," in fact, is merely a convenient way of collecting events into bundles... "Substance," in a word, is a metaphysical mistake, due to transference to the world-structure of the structure of sentences composed of a subject and a predicate" [99].

Indeed, we have no other way to describe an object with a language than by using its properties revealed by the decomposition of reality. However, an attempt to describe a subject *via* his/her properties immediately reduces the subject to an object so that we include one category into the other instead of getting into a "subject–object" opposition. On the other hand, exclusion of the opposition deprives the term "object" of its content and it ceases to function as a sign since it differentiates nothing at all.[2] Its use in semiosis can give birth to any logical contradictions. In fact, Russell is the author of the paradoxes of this kind: Is the set of all subsets that are not members of themselves the member of this set? Currently, the semiotic studies commenced by Charles Peirce, Charles Morris, Alfred Tarski, and others have acquired great importance associated with the analysis of the paradoxes in the theory of sets. Syntactic and semantic analyses have acquired crucial importance in the new foundations of mathematics proposed by the adherers of formalism, intuitionism, and constructionism. The solution of Russell's paradox that he himself proposed is based on his theory implying the existence of a hierarchy of properties and propositions of different logical types. However, this "systems approach" was rejected by many mathematicians. It is evident that independently of any proposed hierarchy of properties, the categories of "entity" type still cannot be confined to them.

Thus, the subject is unidentifiable according to the properties and even cannot be used in plural merely because we actually have *no tool* to distinguish between "subjects" (unlike individuals). In fact, it is impossible to concurrently imagine something as a unity and a set. These are mutually exclusive descriptions of reality although they complement each other. Russell is right: *substance* (similar to *subject,* the principle of the unity of reality) from the standpoint of object-based reality is *nothing.* Let us consider the consequences of this statement in more detail.

All processes studied by science are eventually modeled in a sign-based form in terms of isolated closed systems. The resulting solutions contain time as a parameter and make it possible to "predict" the pattern of a certain process over the entire time scale. *The content, i.e.,* the principles that determine the form of the process $F(t)$, does not change. Actually, this is a *determined future,* which does not interfere with causality. If we regard an object-based representation as the overall content representable and describable in finite sign-based theories, than a subject-based representation characterizes *a change in the content of the process itself* or **evolution.** Naturally, this can take place only in an open system, where the classical research methodology is inappropriate. It is evident that the causes underlying a change in the content "are not contained" in the subsystem itself but are rather determined by the goals of a "suprasystem"; correspondingly, the *future* of a subsystem in the situation of evolutionary changes is not determined by its *past* and only new "future content" of this subsystem is able to explain the patterns of the processes changed by the evolution within this subsystem. However, this *fundamentally unpredictable future* is equivalent to "nothing" for this subsystem itself. As has been noted, the introduction of a teleological description necessarily leads to the violation of the causality principle within any subsystem. In this case, a subject-based reality is the indeterministic potentialities of any system. In this reality, the future determines the present.

Quantum mechanics encountered this very situation when a subject in an indeterministic manner determined the present in our reality. Although the wave function predicts the possible variants of states (or trajectory) in a quite deterministic way, yet the choice of a particular state of the system is determined only in the event of the subject's perception (or measurement). The explanation of wave function collapse by "the interaction with a classical device" does not fit since the device itself in terms of quantum description is also a quantum system. Thus, the interaction between the object and device is described by a new wave function and no reduction takes place. The final stage in a measurement is only the very event of perception of this system by the subject. Von Neumann considered this situation in most detail [2].

Note that it is possible in principle to avoid the troubles of "psychophysical parallelism". For instance, Russell believed that once the psychic sensations are the source of data in physics as well, any scientific knowledge is reducible to psychological knowledge (at least, what takes place in the world is much closer to a psychological explanation). By the way, this also eliminates a number of problems associated with stratification in the description of systems. Psychosemantics implements this approach.

Psychosemantics relies on that the "external" reality is represented to the subject in a certain specific form or, according to Pavlov, the *first signal system*—a psychic reflection of "external reality". Actually, this is a *mental map* or a specific *model* of the surrounding world "beyond" our sensations. An individual relates to this map all his\her actions aimed at satisfaction of the relevant needs. Evidently, the more adequately and the more completely the mental map reflects the "external reality", the more successful the individual adapts to the environment and acts there. Note for clarity that the mental map here is regarded exclusively as a ***result*** of a psychic process of "reflecting" reality rather than the corresponding processes (perception, sensation, thinking, memory, and so on). We also omit *how* a subject obtains knowledge about the "external" world. The only important point here is that such knowledge in the process of adaptations to this external world is representable as a mental map, *i.e., a certain virtual model of this world.* An increase in the complexity of the conditions for existence and the development of needs necessarily lead to the distinguishing of new secondary properties, more complex in their semantics, that are required for the description of the world and navigation in it.

A similarity between the classes of sensations, needs, conditions of existence, and mechanisms of adaptation to the environment creates the basic prerequisites to the generic similarity of the mental maps and, consequently, potentialities for intraspecific communication, which reached its maximal development in humans and provided emergence of the **second signal system** or **language.** Thus, the *second signal* system according to Pavlov is a *translation of the first signal system* (*internal*, intended for one individual alone) *into a sign-based representation* (*external*, oriented towards others). In fact, science, which utilizes a verbal representation of reality, investigates not the world itself but rather *its mental model, i.e.,* our concept of the world, and different branches of science use their own sign systems and notion systems for this purpose. Anyway, all these branches implement a *linguistic simulation of our concept* of Reality since the common tool for describing any phenomenon *in all sciences* is *language.* Thus, it is the semiotic axioms and rules that eventually determine what can be described in theory. According to Heisenberg, the use of classical notions is eventually the result of the total spiritual development of humankind. The concepts of classical physics are fine-tuned concepts of our everyday life and form the most important component of language, which is the prerequisite for the overall natural science. "We have to remember that what we observe is not nature herself, but nature exposed to our method of questioning. Our scientific work in physics consists in asking questions about nature in the language that we possess and trying to get an answer from experiment by the means that are at our disposal" [21].

An interesting approach to analysis of the boundaries of a theoretical (formal) study of reality as an open system is described by Penrose, a well-known mathematician and theoretical physicist, in his last books *The Emperor's New Mind* and *Shadows of the Mind*. He poses the question of whether it is possible to reduce the function of our mind, including realization and understanding, to a certain formal algorithm. In other words, is it possible to construct a theory of the Mind and Consciousness? Let us brief his reasoning and the historical background of this problem.

In 1900, Hilbert formulated the problem, known as Hilbert's tenth problem: to find a computational procedure for deciding whether for a given system of Diophantine equations[3] they have any common solution. Thus, the problem was to find a certain universal algorithm. The first step to its solution was made by Alan Turing in 1936, when he proposed his own particular definition of what an algorithm is, in terms of his Turing machines. However, Yuri Matiyasevich, a Russian mathematician, in 1970 showed that there could be no algorithm that systematically decides yes/no to the question of whether a system of Diophantine equations has a solution. A class of such problems was later discovered. Penrose, having analyzed these problems, concludes that there is a clear difference between determinism and computability. "There are completely deterministic universe models, with clear-cut rules of evolution, that are impossible to simulate computationally" [7].

Then he analyzes the aspect of conscious act, such as *understanding*. First, Penrose considers the line of argument—the well-known "Chinese Room" of the philosopher John Searle—which demonstrates that understanding as an attribute of thinking is irreducible exclusively to computational algorithms [106]. Then Penrose gives a strict mathematical proof based on famous Gödel's incompleteness theorem[4] and algorithmic Turing machine. Note the following statement proved by Gödel: no *formal system* of sound mathematical rules of proof can ever suffice, even in principle, to establish all the true propositions of ordinary arithmetic. Penrose demonstrates that it also follows from this statement that the human intuition and understanding is irreducible to any set of rules. In other word, "We must consider that ... mathematical understanding might be the result of some algorithm that is unsound or unknowable, or possibly sound and knowable but not knowably sound" [7].

Thus, Penrose in the first part of his *Shadows of the Mind* comes to the conclusion that an intricate organization of the brain alone is insufficient for consciousness to appear; the brain must be organized to provide "non-computable" physical processes (the principle of psychophysical parallelism). Then, Penrose assumes, "...the only minds of which we have direct knowledge are those intimately associated with particular physical objects—*brains*—and differences in states of mind seem to be clearly associated with differences in the physical states of brains. Even the *mental* states of *consciousness* seem to be associated with certain specific types of physical activity taking place within the brain" [7]. Interestingly, "Gödel appears to have taken it as evidence that the physical brain must itself behave computationally, but that the mind is something beyond the brain, so that the mind's action is not constrained to behave according to the computational laws that he believed must control the physical brain's behavior" [7].

When analyzing various physical theories and phenomena, Penrose concludes that the procedure of wave function *reduction* (procedure **R**) during measurement unlike its *evolution* (procedure **U**) is a *real non-computable physical process*. Unlike Eugene Wigner, a well-known physicist, who believed that the unconscious matter evolves in a computable fashion but as soon as a conscious entity becomes physically entangled with the state of a system, then something new comes in and a certain physically non-computable process is switched to reduce the quantum system to perceivable reality, Penrose tries to *substantiate* Consciousness by this process. In other words, Wigner *substantiates* this process by the *consciousness of a living being,* whereas Penrose *substantiates* the consciousness of a living being by *this process.* Interestingly, Penrose in his dispute with Wigner unwittingly also utters another view *directly associating* this non-computable process with Consciousness: "There need be no suggestion, with such a viewpoint, that somehow the conscious entity might be able to "influence" the particular choice that Nature makes at this point" [7]. If we eliminate the word *"entity"*, then the *"particular choice that Nature makes at this point"* actually describes the action of a *"free conscious will"*. At least, this phrase evidently presents it as possessing such attributes although in a form of a "turn of speech" and Subject is in fact identified to Nature.

Recently, Penrose elaborates on the idea that gravitation is involved in this process [56]. He believes that "diverging" possibilities[5] change the spacetime metric and Nature (capitalized!) in order to avoid a contradiction is forced "to decide" in favor of an alternative. However, Nature in this understanding is the Subject, the overall Universum as a whole. The gravitation mechanism itself fails to explain a particular choice but only demonstrates its necessity. It is important that quantum "possibilities" are not mere theoretical abstractions but rather actual reality although not belonging to our subsystem since they indirectly manifest in the effects of wave function interference (for example, electron in two slits).

A systems aspect of evolution is also reflected in the stratification of quantum and classical descriptions of reality. Since a physical interpretation of the "quantum" oscillation is absent, the wave function from the very beginning is determined as a complex function. In general, quantum mechanics is a unique discipline that has learned to solve a wide range of problems without understanding them, as was noted by Einstein, Bohr, Feynman, and others. Many researchers (including Louis de Broglie) attempted to reduce the quantum reality to a classical one using "hidden variable theory". The reasons why the hidden variable hypothesis remains attractive for scientists are not only the wish to restore classical determinism in the area of quantum phenomena. This is rather an attempt to preserve an "objective" (yet object-based) "depersonalized" description language,[6] which excludes the subject from physical theory. The example of Bell, who demonstrated that this is an unfeasible solution, is another confirmation.

Explaining the motives that induced him to study the issue of hidden parameters in quantum mechanics, Bell names as the main factor his dissatisfaction with that the generally accepted interpretation of quantum mechanics relies on the decomposition of the physical world into two parts of classical and quantum phenomena, described in a fundamentally different fashions; moreover, any clear explication of their relation is absent. Bell says that the most comprehensive description of the world's state in general or any of its parts is $(\lambda_1, \lambda_2, ..., \lambda_i; \Psi_1, \Psi_2, ..., \Psi_i)$, where λ_i are the classical variables describing the state of an experimental device (positions of switches, pointers, and so on) and Ψ_i are the corresponding quantum mechanical functions. This nonuniformity or, more precisely, dualistic description implies the existence of a certain boundary that separates the areas of classical and quantum phenomena or, in more general terms of philosophy, the boundary between the cognizing subject existing in the classical world, adapted to it, and "equipped" with the corresponding system of notions and tools and the quantum mechanical object that he cognizes. Even assuming that there is a certain agreement in that, at least, the switches and pointers are in the classical world, "on our side" of the boundary, the opinions on the "depth of its occurrence" are most different.

Presumably, the quantum world is in fact a different reality where an imaginary unit appropriately marks the properties hidden from our perception.

Note that Schrödinger's equation $\frac{\partial \psi}{\partial t} = \frac{ih}{2m} \Delta \psi$ can be written as $(i\partial_t + \partial_{xx})\psi(x, t) = 0$, which is analogous to the diffusion equation $(\partial_t + \partial_{xx})u = 0$. Thus, Schrödinger's equation is derived from the diffusion equation by a simple replacement of $t \to it$.

Correspondingly, whereas equation $(\partial_t + \partial_{xx})u = 0$ has exponential solutions, the replacement $t \to it$ results in harmonic ones. Naturally, interference and diffraction are the main phenomena of this "imaginary" world. Time (ours) has no sense in a spectral representation since the overall process in this case is represented as a whole (unfolded in space and time since the harmonic has no spatial constraints[7]). "Our time" as a *variation* is generated by how we look at this solution through a time window (or a spectral window equivalent to it), that is, the cause lies in the uncompleted spectral transformation. The indeterminacy relation is derivable from the known relation: the product of momentum spectrum width (spectral window) and its duration is a constant. It is evident that the time under the "Florensky plane" is different. This time changes the spectral composition (type of representation), which is equivalent to the change in the *content* of process, and the latter in our world is instantaneous (unlike the form). Actually, there is no notion of velocity in our sense of the "imaginary world". For example, Feynman diagrams show that the time may go forward and backward (in our reality, this is equivalent to the particle–antiparticle transition) and "information" is transferred with an infinite velocity (this is the only way how a particle can "probe" all possible pathways and, indeed, the backward time "flow" also suggests faster-than-light velocities). In general, only x and it have sense there. In fact, it is possible to imagine another world "under" our space—the world of potentialities. Only its real component, *i.e.*, the moduli, which have the meaning of probabilities, has any sense for our consciousness as a "subsystem". In this world, ψ function defines an integral situation rather than an object. Conceivably, this and that worlds are interpretable from the standpoint of psychology as consciousness and the unconscious. The experiments on quantum teleportation also suggest that the Universum as a whole really exists.

If we include the teleological principle into a systems-based description to a full degree, it is then necessary to include the subject as well and to separate the notions of activity and reactivity. "Presumably, the substitution of activity and purposefulness for reactivity was and is determined by that the use of natural science and experimental methods in general is as a rule combined with a causal explanation of behavior. This explanation is traditionally related to the paradigm of reactivity, whereas the paradigm of activity and goal orientation is related to a teleological explanation" [98].

Thus, the representation of Universum as an open system must take into account two sides of the description, namely, the object-based (multiplicity) and subject-based (unity) aspects. As is mentioned above, the desire of natural sciences to retain an "objective" (in fact, object-based) "depersonalized" description language, excluding the subject from its paradigm, fails to achieve the goal. The complementarity of subject-based and object-based descriptions of systems means that a causal explanation requires consideration of isolated systems and a teleological explanation, of open ones. The subject here is regarded as an integrated aspect of the system (actually, Universum). In this sense, procedures **U** are defined according to Penrose in an object-based paradigm and procedures **R,** in a subject-based one. Evolution as a creative event that changes the content of a certain process interferes with cause-effect relations since the mind is presented with the effect (new content) but it cannot distinguish or find its cause in its past (old content). *The process of Evolution can be imagined as creation of a hierarchical structure of Subject's (observer's) reference frame with increasing complexity:* physical → biological → psychic → social and so on. The emergence of each observer's reference frame is accompanied by emergence of new more flexible qualities and new principles (objective vectors) within each observer's reference frame. The new principles determine the rules for the behavior of objects in each observer's reference frame and a decrease in the rigidity of new qualities increases the rate of evolutionary processes. A condition for theoretical justification of an observer's reference frame is *its origin from a single state.* For example, the overall physical Universe, which originated from one singularity, is undoubtedly one of the basic observer's reference frames[8] of the unified meta-subject. It has its own principles, which are universal with regard to the subsequent observer's reference frames. The properties of each previous level form the grounds and represent the condition for the existence of the properties of the next level (physical → biological → psychic → social and so on). We will match the corresponding subject to each observer's reference frame(keeping in mind that this is a unified meta-subject represented in this particular observer's reference frame). In biology, a family, genus, species, and individual are examples of different observer's reference frames since they all originate from a cell (zygote) and a common ancestor as well as all cells of an organism have the same marker - a specific set of chromosomes.

In this perspective, evolution may be regarded as the process by which the meta-subject creates the necessary conditions for *implementing its "needs", which in an indirect manner reveal the content of evolution itself.* The sequence of creative events is the creative process in its essence. The content of each stage of the evolution is *the condition for implementation of its next stage,* which reveals *the meaning of its previous stage* (which completely fits the postulates of the systems theory [107].

The diversity of evolutionary processes (cosmological, physical, chemical, biological, psychological, social, *etc.*) forms a hierarchy of the corresponding "subjects". The hierarchy of evolutionary processes generated by the principles, objective functions, or "needs" of the "subjects of evolution" of different levels allows us to successively reveal the conceptual component of Development at any of its stages. Nonfiniteness of the content of Development in this case is evident: each stage is determined by the entire infinite chain of events of infinite *redefinition of the meaning of "subject".*[9] If the number of observer's reference frames (which is determined by the final content) were *finite,* the eventual content (definition) of Development and Subject would also be limited, while the Subject would be reduced to object and Development, to mechanical movement. Thus, the stages of Development are concretized in evolutionary processes, which can be regarded as the stages in the *concretization of subject-based space* and the emergence of new qualities (physical, psychological, social, and so on). These qualities are implemented in new observer's reference frames that determine *concrete subjective realities*. In other words, ***the Subject as the opposition to object-based reality has always existed,*** and we are its "representatives". Therefore, it was neither a mechanical world nor some mechanical fluke that gave rise to Development with its biological observer's reference frame, and organism, *but vice versa.* A mechanical system will always remain mechanical because it is always "closed" by definition with respect to its eventual content. Two types of processes are concurrently going on in the world: one is associated with a growth in entropy (mechanistic) and the other, with its decrease (evolutionary); entropy is indeterminable in open systems. Here, semiotics demands equality too. For example, our Universe according to astrophysical observations emerged from a fluctuation of the vacuum. *Consequently, all quantitative changes accumulated in it in total must be zero at any time moment of its development* (including the equality between the entropic and negentropic processes). This implies a direct *necessity of evolutionary processes* for the very existence of *physical* Universe.

Any sign-based method of description is limited by definition[10] and makes it possible only to successively reveal in the finite notions a certain "nonfinite cause" of creativity, which, similar to the classical science, goes to "bad infinity". Whereas the infinity of cognition in classical science is determined by infinite diversity of objects, this infinity in a subject-based approach results from the diversity of "subjects" of different concretization levels. In this case, the goal of the subject-based analysis is the revelation of the meaning of each creative event in the evolutionary process *via* the system of goals ("values") of the "subject" at each concretization level rather than the endless process of mechanistic comprehension of the "eventual contents" of the world as an open system *via* the meanings (notions). Science interprets these goals as principles, such as the least action principle, natural selection principle, and so on, implicitly (or shamefacedly?) "anthropomorphizing" "sluggish Nature" and ascribing to it some goals and purposefulness.

Thus, the mechanistic and evolutionary processes must be distinguished. The evolutionary process is implemented in a sequence of creative events (atemporal "jumps") that generate the eventual content. The mechanistic process is the mode of existence of this eventual content in an object-based spacetime, which is implemented in a certain successive change in the forms (for example, in the intensities of the object's properties) that manifest this content.

A mental map appears and is fine-tuned as an image of the conditions necessary to attain and implement the needs as a result of subjective activity within a particular subject-based reality. The conditions are represented in a mental map as the tools necessary for implementing the needs, which we refer to as objects. These conditions are achieved *via* an action (activities). The mental map is our *understanding* of the "external world" and is constructed based on the components of the overall flow of sensations that are independent of us. The direct and lasting experience of the set of these sensations is related to an object on the mental map. This is *an objective portion of our experiences* since it is not directly connected with our will. As for the sets of sensations that are directly connected with the volitional functions, an individual consciousness[11] associates them with itself and determines them as *a subjective portion of our reality.* The set of sensations this is not directly associated with the act of will is interpreted as *external* relative to the subject, related to with the source of these sensations, and related to an "external" object of the "external" "object-based reality".

The conditions for self-perception and understanding of the world comprise not only the objective principles of subject-based Reality, but also *"genetically specified" elements of the mental map,* which we understand as congenital reflexes. They form a ready-to-use link (a pattern to be used in order to satisfy certain needs under certain conditions) between the classes of sensations. The "internal" tools for satisfying the needs are defined as instincts and represent the primary tools in implementation of the needs.

If a subject utilizes only instinctive resources for satisfying its needs, this type of satisfaction is referred to as *direct.* If the satisfaction of a need requires creation of certain *preconditions,* this type of satisfaction can be regarded as *instrumental.* The instrument in this sense acts as a *mean for creation of preconditions for direct satisfaction of a need.* Thus, the means for creation of instruments of different levels, both physical and social, emerge on the basis of mental construction. In fact, the reason and the mental map itself are such instruments too.

We can trace in detail how a stimulation emerges, spreads, and transforms when an organism interacts with the environment. However, the subject does not have and cannot have any "sensory organs" that *transform stimulation into a sensation* or *physical into a psychic.* A canonical approach "cocoons" us in sensations, which in our view are the markers of our absolutely inconceivable "interaction" with the "real external object-based world" as the source of these sensations. As for the only directly given existence—subjective reality, where we live—we are forced to declare it "virtual", secondary (derived), and seeming. Psychosemantics relies on the opposite statement: the subject constructs an object through the lens of its needs and goals, while the object-based reality is a derivative with respect to our subjective reality. The question of wherefrom the images actually come to our consciousness is in its essence the question of religion rather than science.

We determine whether a mental model of the world (map) is adequate by the degree to which we feel at home in this world (out own subject-based reality), *i.e.*, to the degree to which we *can efficiently reach out goals and satisfy our needs - this is the only thing that matters.* In terms of canonical approach, this demonstrates the degree of our fitness to the "external source" ("environment", "surroundings", and so on). According to psychosemantics, science investigates the objective (common for all) components of mental maps rather than the world itself.

The meanings of objects in the second signal system are determined by the overall mentality rather than a single individual. Since mentality describes the essential components of a subject-based aspect as the totality of subjective realities, the meanings of objects must be revealed in a probabilistic manner taking into account their possible interpretations in this totality. Thus, mentality includes all possible conceptual interpretations of the common elements in mental maps.

The "subject" itself can be represented in the mental map only as an individual—*an observer's reference frame, i.e.*, a hierarchical system with certain sets of properties (organism, temperament, character, identity, and so on). In this context, other "subjects" represent other possible mental systems. *Each subjective reality is interpreted as a "view" of the Subject on Reality from its local point or its own reference frame.* Any evolving system may represent the Subject's (observer's) reference frame with functional capacities of different levels. The transition from one subjective reality to another is associated with the transformation of the corresponding "system of coordinates". All observer's reference frames *equivalent in their functional capacities* and in general mechanistically reducible to each other with the help of a certain class of transformations[12] correspond to *equivalent* subjective realities that together form *subject-based reality.* The rules of these transformations are also objective and reflect the quantitative transformations within the corresponding subject-based reality.

Thus, the "subjects" (in classical understanding) appear on the mental map as a result of *conditional attribution.* An element of the mental map is recognized as equivalent to a certain *other observer's reference frame* of the subject and is represented as a *new subjective reality.* In this case, the fact that an entity belongs to the class of subjects is in essence equivalent to being "ensouled" or "animated", to possess consciousness, and to be able to construct a mental map from this observer's reference frame.[13]

Actually, the Subject is a mere *attribute of the unity of the world, i.e.*, of any subjective or subject-based reality. The classical science first "tears" the Universum into objects representing it as a "wasteyard" of separate things and then makes a long and painful trip of proving the relation between all and everything by introducing forces, fields, and all other kinds of metaphors. Using such an approach, it is really difficult to understand the paradoxes of Reality, such as the EPR paradox, since this is a manifestation of the *unity* of Universum.

The notion of *objectivity* is understandable *via* the initial *general principles in organization* of subject-based Reality rather than *via* referring to the initial existence of *external* "object-based Reality". By comprehending its subjective reality in terms based on the *objective principles specified by Evolution,* the mind erroneously "alienates" this reality from itself as an *object-based entity.* Thus, we see that an *objective* reality is implicitly replaced by an *object-based* one.[14]

Any notion has *meaning* and exists not by itself but rather *in relation of the defined object to the need of the subject who defines it.* The *existential quantifier* itself is the certification by a subject that a certain phenomenon (object) *is represented in at least one individual consciousness* (or in a particular class of observer's reference frames). In essence, *Consciousness is an integral attribute of the Subject itself whatever constraints are imposed* (in any observer's reference frame). In fact, consciousness ascertains the "presence" of a *"perceiving subject" of a certain concretization level* in a particular observer's reference frame. Then the phrase "an object exists" is a mere statement (*certification*) *of the perceiving subject* that this phenomenon is represented in a particular subject-based Reality or *is comprehensible to Consciousness under this particular constraint.* In an analogous manner, *existence* may be defined relative to a subject-based aspect as a whole. This means that *something* must be "present" in all mental maps resulting from all permitted transformations of these observer's reference frames (physical, biological, social, and so on). Consequently, *the boundaries of individual consciousness* coincide with *the objective constraints of the subject's (observer's) reference frame.* Thus, it is not at all necessary to search for other worlds somewhere beyond our Universe. Our sensations are not born from a vague "interaction" with a hypothetical "object-based world" but are rather a peculiar kind of projections of the previous concretization layers (infinite?) of subject-based Reality. The properties of quarks or spinors are inaccessible to our sensations but they particularly determine the physical sensations of "macroworld" accessible to us. Thus, the subject-based reality is objective in its nature and needs no postulation of a "transcendent", inaccessible, and, in essence, ideal (in Platonic sense) world of objects.

Separation of any notion from the motives and needs of a subject who has generated this notion actually deprives it of its *meaning* and makes any term a mere indicator of property relations or a *value*. *A notion* can be defined only in terms of its *value for a subject*. If science tore off the meaning from a notion or introduced senseless notions, it would degenerate into an abstract and useless game with toy blocks (values) rather than construct anything *sense-bearing* (useful and *necessary* for a subject). At least, Hilbert failed to achieve his dream to finally formalize mathematics using only finite methods in the research into theories and to completely replace semantics with syntax. This program had emerged to be unfeasible, as was later demonstrated by Gödel in his incompleteness theorem.

The classical science implicitly ascribes real existence to the notions themselves as a "material" manifestation of a "virtual" world, which "exists" without the subject itself.[15]Correspondingly, the notions themselves replace the direct reality, which can be nothing but subject-based.

Comprehension is reduced to the point that the boundary (definition) of subjective reality changes in the course of evolution. Objective phenomena, "external" with respect to older constraints of the observer's reference frame, become accessible to consciousness under *new constraints*.[16] Then, what determines an individual consciousness "from outside" can become accessible to it as the comprehension "from inside" and as an object-based earlier incomprehensible principle of its determination (a new law). The objectivity of subjective realities is a consequence of the fact that *they all are different constraints of the single subject-based Reality*.

Since the mental map is our subjective representation of Reality (its virtual mental model that we use to navigate the surrounding world), the understanding of the psychic laws and behavior of an individual requires that we learn how to construct at least the corresponding fragments of this map. This reasoning necessarily suggests the following strategy for constructing the mental map:

1. The theoretical equivalent of a mental map (concept) must provide *a sign-based representation adequate to this map* in a certain space of characters. Such space of characters (qualities) must ensure the possibility for a *full description* of phenomena and
2. The formal semiotic method for representation of objects in a mental map must be *universal, i.e.,* should be applicable to the objects of any nature in both natural and humanitarian sciences.

An ordinary description (definition) of any object (phenomenon) in a language is implemented by *listing its qualities and their intensities* (the degree of manifestation of these qualities). Correspondingly, a formal semiotic definition of object Ω, an element of the mental map implemented in the first signal system, can the object Ω is equivalent to the vector representation $\bar{U} = \{Q_1, Q_2, ..., Q_j, ..., Q_n\}$,[17] where coordinate Q_j is the intensity of the *j*th quality (j = 1, 2, ..., *n*). Thus, any object, an element of the mental map implemented in the first signal system, can be uniquely related in the concept (language or second signal system) to a certain vector $\bar{U}$ in the space of qualities, *i.e.*, $\Omega \rightarrow \overset{\cdot}{U}$.

It is evident that the individuals living under similar conditions and having similar relevant needs, *i.e.*, the individuals of the same mentality, form a common semiotic system of communications.[18] In addition, all interpretations of meanings for the objects within a uniform mentality must be implemented with a *certain probability*.[19] This is associated with that *actualization of a particular interpretation depending on the context of particular situations the frequency of which is determined by the general conditions of life and surroundings of these individuals.* Different conditions interfere with the adequacy of communications (the mutual understanding of individuals). As an example, recollect the behavior of Eliza Doolittle, a cockney flower girl, in the high society environment of Professor Higgins in *Pygmalion* by Bernard Show. A diversity of interpretations of any notion is well known to linguists [108] and is in part represented in thesauri. They usually inform about a particular field where a particular meaning is relevant (for example, *everyday usage, dialect, or science*). Any attempt to describe an object as a "self-sufficient" and *independent of the conditions of its perception* inevitably makes it polysemous and "virtual".

This is Heisenberg's line of reasoning concerning the structures of logic and language in connection with the paradoxes of quantum mechanics:

"These other structures may arise from associations between certain meanings of words; for instance, a secondary meaning of a word which passes only vaguely through the mind when the word is heard may contribute essentially to the content of a sentence. The fact that every word may cause many only half conscious movements in our mind can be used to represent some part of reality in the language much more clearly than by the use of logical patterns" [21].

Let us construct a formal algorithm for the transition from the set of subjective interpretations of an object by individuals to *general linguistic meanings that objectively describe the object via* its states within a given mentality.

Let the matrices $\left(u_{qj}^{(i)}\right)$ be descriptions of L objects ($i = 1, 2, ..., L$) by each of the N respondents ($q = 1, 2, ..., N$). Each row of the matrix is the description the ith object by the qth respondent or a unit vector that determines the ratio of semantic components in this description:

$$\vec{u}_q^{(i)} = \frac{\vec{U}_q^{(i)}}{\left|\vec{U}_q^{(i)}\right|},$$

where $\left|U_q^{(i)}\right| = M_q^{(i)}$ is the length of a nonstandardized vector describing the object.

A respondent describes the object with the help of n descriptor scales ($j = 1, 2, ... n$) and the results of this description are represented in an n-dimensional semantic space. Actually, this is analogous to N independent sets of measurements of L phenomena (situations or objects) according to n characteristics.

Thus, the results can be presented as L matrices that are the descriptions of L objects by N respondents, *i.e.*, each matrix contains N descriptions of only one of the L objects:

$$\left(u_{qj}^{(i)}\right) = \begin{pmatrix} u_{11}^{(i)} & u_{12}^{(i)} & \cdots & u_{1(n-1)}^{(i)} & u_{1n}^{(i)} \\ u_{21}^{(i)} & u_{22}^{(i)} & \cdots & u_{2(n-1)}^{(i)} & u_{2n}^{(i)} \\ \cdots & \cdots & \cdots & \cdots & \cdots \\ u_{N1}^{(i)} & u_{N2}^{(i)} & \cdots & u_{N(n-1)}^{(i)} & u_{Nn}^{(i)} \end{pmatrix}.$$

The rows of the ith matrix are the vector descriptions of the ith object by the qth respondent:

$$u_q^{(i)} = \left\{ u_{q1}^{(i)}, u_{q2}^{(i)}, ..., u_{qj}^{(i)}, ..., u_{q(n-1)}^{(i)}, u_{qn}^{(i)} \right\}$$

Evidently, these vectors in a *particular mentality* form rather narrow "bundles" of agreed opinions[20] on each of the objects in the n-dimensional *semantic space;* they can be associated to the unit vectors[21] $\vec{s}_m^{(i)}$ ($m = 1, 2, ..., K$), where K is the number of "bundles" of the vectors of individual estimates (Fig. **7.1**). Each vector $\vec{s}_m^{(i)}$ specifies a certain *meaning* of the ith object common for the overall mentality (its permissible interpretation that is common for all individuals).

Thus, we replace *N individual estimates* of each object with *m (K) agreed interpretations* of this object.

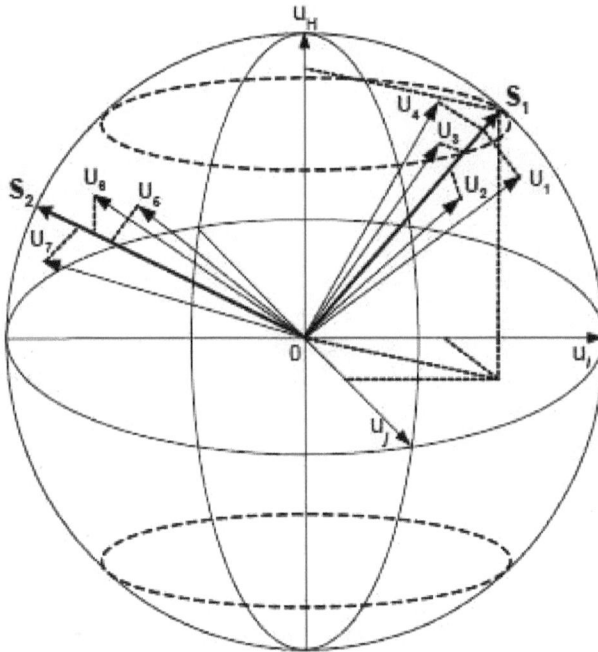

Fig. (7.1). Definition of the states (*S*) of an object according to estimates (*U*); Fig. 3.9 in vol. 1.

The set of *K* common *agreed* stable *interpretations* (or "*meanings*" in linguistics) of each object in the studied uniform mentality[22] shall be referred to as independent *states of the object* $\vec{s}_m^{(i)}$, (*m* = 1, 2, ..., *K*).

This allows us to calculate the *main directions* along which the vectors $\vec{u}_q^{(i)}$ form the bundles. Since the direction of our interest must pass through the center of density of this bundle, the vectors that form it give the largest projections onto this particular direction. The bundle density is determined by the sum of squares[23] of the projections in this direction. We associate the unit vectors $\vec{s}_m^{(i)}$, mentioned above, to these bundles. Evidently, they characterize *the most expected interpretations of the ith object* (phenomenon) by this mentality.

The search for $\vec{s}_m^{(i)}$ is started from solving the standard characteristic equation[24]

$$\left(A^{(i)}\right)\cdot\vec{\phi}^{(i)} = \lambda^{(i)}\cdot\vec{\phi}^{(i)} \tag{1}$$

note that all $\vec{\phi}_m^{(i)} = \left(c_1^{(i)} \cdot \vec{u}_1^{(i)} + c_2^{(i)} \cdot \vec{u}_2^{(i)} + \cdots + c_N^{(i)} \cdot \vec{u}_N^{(i)} \right)$ the eigenvectors, and $\lambda_m^{(i)}$, the eigenvalues of the matrix $(A^{(i)})$, which are equal to the sum of the squared projections of vectors $\vec{u}_q^{(i)}$ onto the directions defined by vectors $\vec{\phi}_m^{(i)}$ [109]. If the matrix is symmetric, all eigenvalues are positive. Actually, $\vec{\phi}_m^{(i)}$ are the orthogonal basis of the vector space defined by matrix $(A^{(i)})$, while K coincides with the rank of this matrix (the number of linearly independent vectors).

Thus, the matrix $(\Phi^{(i)})$ of the eigenvectors $\vec{\phi}_m^{(i)}$ defines the principal directions or *states of the ith object in the space of objects:*

$$\left(\Phi^{(i)} \right) = \begin{pmatrix} c_{11}^{(i)} & c_{21}^{(i)} & \cdots & c_{K1}^{(i)} \\ c_{12}^{(i)} & c_{22}^{(i)} & \cdots & c_{K2}^{(i)} \\ \cdots & \cdots & \cdots & \cdots \\ c_{1N}^{(i)} & c_{2N}^{(i)} & \cdots & c_{KN}^{(i)} \end{pmatrix}$$

The total sum of the squared projections of all unit vectors onto all directions is actually the total sum of their squared lengths and is equal to their number, *i.e.*, N.

Then,

$$\left(\phi_m^{(i)} \right)^2 = \sum_{q=1}^{N} \left(c_{mq}^{(i)} \right)^2 = \lambda_m^{(i)}$$ as is mentioned earlier, and

$$\sum_{f=1}^{K} \lambda_m^{(i)} = N, \tag{2}$$

where Eq. (7.2) determines the sum of squares of all unit vector observations represented in the basis of eigenvectors $\vec{\phi}_m^{(i)}$

The following relation can be used to find the *semantic representation of the object's state* $\vec{s}_m^{(i)} = s_{m1}^{(i)} \cdot u_1 + s_{m2}^{(i)} \cdot u_2 + \cdots + s_{mn}^{(i)} \cdot u_n$, ($\vec{u}_j$ are the basis vectors of the semantic space). *i.e.*, the representation of vector states *in a semantic space* rather than in an *object-based space:*

$$c_{mq}^{(i)} = \vec{u}_q^{(i)} \cdot \vec{s}_m^{(i)} \tag{3}$$

which defines the projection $c_{mq}^{(i)}$ of the qth observation of the ith object $(u_q^{(i)})$ onto the mth principal component $\vec{s}_m^{(i)}$.

Note that the relation

$$\lambda_m^{(i)}/N = P_m^{(i)} \tag{4}$$

gives the share of the variance for the total sample of estimates for the ith object that corresponds to the mth state [109] or the *probability of the mth interpretation of the ith object* defined by the state $\vec{s}_m^{(i)}$. It is commonly assumed that $\lambda_1 \geq \lambda_2 \geq \cdots \geq \lambda_K$ In principle, characteristic Eq. (7.1) may have N solutions; however, it is no use considering the states with the eigenvalue smaller than the measurement error; thus, $K \leq N$.

In the case of a total population (infinite experiment), $N \to \infty$ and the eigenvectors $\vec{\phi}_m^{(i)}$ in Hilbert space **H** pass into the Eigenfunctions $A \cdot \phi_m^{(i)}(\bar{u}) = \lambda_m^{(i)} \cdot \phi_m^{(i)}(\bar{u})$ of operator $\hat{A}$.

Then,

$$A \cdot \phi_m^{(i)}(\bar{u}) = \lambda_m^{(i)} \cdot \phi_m^{(i)}(\bar{u})$$

.

Defining $\vec{\phi}_m'^{(i)} = \vec{\phi}_m^{(i)}/\sqrt{N}$, we have $\vec{\phi}_m'^{(i)} = \vec{\phi}_m^{(i)}/\sqrt{N}$ and $\int_V \left[\vec{\phi}_m'^{(i)}(\bar{u})\right]^2 d\bar{u} = P_m^{(i)}$.

The last equation determines the probability for detection of the ith object in state m (the integral is taken over the whole configuration space V). In other words, $P_m^{(i)}$ is the *probability to get the interpretation* $s_m^{(i)}$ *of the object* $\bar{u}^{(i)}$ *in the studied mentality.*

In a classical case, the ith object is defined by a single state with a probability $P^{(i)} = 1$. If there are several states, especially, if the conditions for observing them are necessarily alternative[25] (for example, the observation conditions with the descriptors "leadership" and "submission" in psychology), it is still possible to associate the ith object to a vector; in this case, this vector is the superposition of all states defined by the vectors $\vec{s}_m^{(i)}$.

A method to overcome redundancy in descriptions of objects is the transition from basic properties (and the corresponding semantic units) to the orthogonal factors uniting synonymic components of a description in a given class of objects; this considerably decreases the dimensionality of semantic space. In essence, this is the separation of group factors by using the method of principal components applied to semantic units [52]. Naturally, in this process, *both **the lengths of semantic vectors and their angular relations**,which define the objects belonging to a certain class,**must remain unchanged*** since we *describe* just *the same system using other notions.*

Since each class of objects is associated with certain goals of an individual (needs and the conditions for their satisfaction), the *found factors reflect the psychological, social, and other attitudes of perception* and *estimates* of the objects belonging to a given class. They determine *the stable elementary gestalts of perception of this class of objects* for a particular mentality. It is evident that other perception attitudes correspond to another class of objects (or other mentalities). Since each factor determines a certain regression equation [111] that in a particular way relates the properties for a given class of objects, we automatically get *the specific laws for relation of properties* in this class.

Thus, we implement the representation of the states of objects in an *r*-dimensional (r ≤ n, where *n* is the number of basic properties) space of independent factors (attitudes):

$$\vec{s}_m^{(i)} = \sum_{j=1}^{r \leq n} \beta_{mj}^{(i)} \cdot \vec{f}_j ,$$

where vectors $\vec{f}_j$ define the orthogonal factor basis of a new semantic space ($j = 1$, 2, …, *m*).

A varimax rotation of the factors, retaining the orthogonality of factor structure [109] simplifies their lexical expression *via* the basic properties and completes the transition from the *factor semantic* space $\left\{\vec{f}_j\right\}$ to *categorical semantic spac*$\left\{\vec{g}_j\right\}$:

$$\vec{s}_m^{(i)} = \sum_{j=1}^{r} \gamma_{mj} \cdot \vec{g}_j \tag{5}$$

Principles for Construction of Motivational Spaces

The direct stimulus for an individual to construct a mental map is the necessity to satisfy the own needs (refill necessary resources) based on a signal (sign-based) representation of Reality. The individual distinguishes a certain pattern from the flux of sensations of various modalities only when this signal is *stably* associated (directly or indirectly) with satisfaction of a certain need. In the mental map, the individual sees it as the object associated with satisfaction of the corresponding need (being the *mean* for satisfaction of this need or a *condition* for it). *The need that made this object to be distinguished as a certain set of sensations determinesits significance for the individual.*

A need as a certain deficiency in resources for the individual's life is always actualized under certain conditions, *i.e.*, in the "environment" of objects represented in the individual's mental map.[26] The need is implemented *via* the most efficient means for its satisfaction that are available to an individual under given conditions. This determines different significances of objects for an individual when certain needs are actualized.[27] Using significance (desirability) of objects for an individual at a current moment, we can calculate the "**motivational vector**" $\vec{z}_l$ ($l = 1, 2, ..., r$) constructed for any rth condition. This vector defines the motivation for the individual to select particular objects under given conditions. The very same vector determines the most relevant significances (states) of the object $\vec{s}_i$ of all the possible ones under given conditions. Actually, the motivational vector together with the states of objects determines the *mental state* of an individual and the corresponding activities aimed at satisfaction of the relevant need under given conditions.

It is evident that the most relevant meanings of objects are those that maximally meet the motivation of individual's activity at a current moment. *The optimal meaning* of an object can be *selected* among all possible interpretations[28]according to the angular proximity of the semantic vector states of the object $\vec{s}_{ij}$ ($j = 1, 2, ..., n$) to motivational vector $\vec{z}_l$. Formally, the relevance of any meaning of an object is definable as a value proportional to the cosine of the angle between the directions of motivational and state vectors, $\cos \theta_{ijk} = \vec{s}_{ij} \cdot \vec{z}_l$. The sign of the scalar product determines the sign of proposition and its value determined the degree to which its use by an individual in the corresponding interpretation is semantically *adequate* under given conditions.

Thus, the motivational vector can be interpreted as the factor that provides *the best balance of the properties* that satisfy a certain need of an individual. A continuous dependence of the relevance of object's states on motivational vector $\vec{z}$ (the conditions for implementation of a need) determines the variation in the significance of an object unlike its meanings. Note that the particular goal for which the object is used reveals its significance (*purpose*) for a subject.

A biologist asked what causes mutual sexual attraction of genders will answer that this is the *need* of Evolution or Mother Nature, which created us. Actually, a physicist will give the same answer to the question why opposite electric charges are attracted: this is imperative or *the need* of Mother Nature. In a formal manner, the *need* is definable as a certain objective principle that is implemented within a certain class of objects and discloses to us the meaning of a particular process[29] (*why* it takes place).

We have defined above the motivational vector as the factor that ensures *the best balance of properties* that satisfy a certain need. An actualized need appearing as a deficiency in resources creates a tension in a system (biological, psychological, social, and so on) and the system tends to compensate this deficiency. Thus, we may regard the need (independently of whose need it is) as satisfied if the interaction leads to a certain strainless stationary state of the system.

In the previous sections, we took N matrices of experiments $\left(u_{aj}^{(i)} \right)$ and got $L \times K$ matrices $\left(s^{(i)} \right)$ that define the directions of K vector states $\left(s^{(i)} \right)$ for each of the L objects (phenomena) in the semantic space of categories $\{ \vec{g}_1, \vec{g}_2, ..., \vec{g}_r \}$ and the probabilities $\left(P_m^{(i)} \right)$ of these states for the studied mentality.

A categorial representation of objects is based on their *significance* rather than their *meaning,* which can be revealed by relating the objects to *motivation,* as a certain target function of the studied mentality that determines their *importance for the subject of the need.*

Evidently, each motive in a particular mentality is determined by a certain need and concrete conditions for its satisfaction, *i.e.,* the motive is related to the corresponding motivational vector $\vec{z}_l$. This vector characterizes the balance of basic properties that is optimal for satisfying the studied need under given conditions. In this case, the "desirability" $R_{lm}^{(i)}$ of the ith object in state m relative to the given motivation $\vec{z}_l$ depends on both the angular proximity of its vector state $\vec{s}_m^{(i)}$ to the direction defined by vector $\vec{z}_l$ and the "length" $M_m^{(i)}$ of vector state $\vec{s}_m^{(i)}$, as well as the probability of this state, $P_m^{(i)}$.

Correspondingly, the eventual desirability of an object $R_l^{(i)}$ (or its rating) is defined by the scalar product of the vectors[30].

$$R_l^{(i)} = k \cdot \vec{U}^{(i)} \cdot \vec{z}_l = k \cdot \sum_{m=1}^{K} P_m^{(i)} \cdot M_m^{(i)} \cdot \left(\vec{s}_m^{(i)} \cdot \vec{z}_l \right) = k \cdot \sum_{m=1}^{K} \left(\frac{\lambda_m^{(i)}}{N} \right) \cdot M_m^{(i)} \cdot \left(\vec{s}_m^{(i)} \cdot \vec{z}_l \right), \qquad (7.6)$$

where $i = 1, 2, ..., L$ and k is a proportionality coefficient associated with the relevance of the need (an analog of charge in physics), which can be determined by subsequent normalization, $|\vec{z}_i| = 1$.

Solving the system of L equations, Eq. (7.6), we get the ideal balance of the object's properties that would satisfy the need in question under given conditions. This balance is defined by *motivational vector* $\vec{z}_l$.

Once knowing, $\vec{z}_l$ it is possible to predict the desirability of any object belonging to a given class according to the sample estimates of its properties under given conditions and for the studied mentality. Thus, we actually get the precise knowledge of the rules according to which any interaction in the studied "subject–object" system is implemented.

Each term $P_m^{(i)} \cdot M_m^{(i)} \cdot \left(\vec{s}_m^{(i)} \cdot \vec{z}_l \right)$ in Eq. (7.6) is the contribution of the corresponding state to desirability (rating) of the object. Replacing Eq. (7.5) into Eq. (7.6), we get the decomposition of object's rating according to categories, which makes it possible to calculate the contribution of each category to the desirability of this object.

Note that the same categories under different conditions can give different contributions to the desirability of objects. A standard statistical approach actually fails to distinguish these regularities.

Further, it is possible to decompose any object in the orthogonalized motivational space $\{\vec{z}_1, \vec{z}_2, ..., \vec{z}_l, ... \vec{z}_r\}$ for the studied mentality. This is the background for analysis of the "subject–object" force[31]interactions, which makes it possible to reveal the meaning and significance of any phenomenon for a particular mentality, *i.e.*, allows for quantitative determination of *what are the needs in a given mentality that this object can satisfy and to what degree the properties of the object are demanded within each need* [51].

Different states of an object reflect its possible potential value as a tool for satisfaction of different needs.

For example, if the need of a body in thermoregulation is not satisfied by its internal resources, the external tools forming the class "clothing" ("fur coat", "jacket", "T-shirt", *etc.*) acquire certain value for a subject. It is evident that their significance for an individual is different in Arctic and in tropics. Correspondingly, the potential significance of any object is revealed in a certain motivational space of a particular need. In addition, clothing is associated not only with a physical need, but also with esthetic, sexual, and/or with social needs, prestige, and so on.

CONCLUSION

Thus, we get the possibility to predict the desirability of objects in satisfying various needs under different conditions as well as to construct motivational spaces for different types of mentalities. This technique makes it possible to approach the theoretical probabilistic modeling and prediction of motivations for behavior of an individual and the mentality in general.

NOTES

[1] Any field of science explicitly or implicitly defines its subject and specific class of the objects it studies (*gases* in gas dynamics; *fluids* in hydrodynamics; *personality* in psychology; goods in marketing; and so on). *It is evident that by separating a certain class of phenomena, we implicitly specify the factor for their selection from the surrounding world* (it is no coincidence that all gases or all fluids are somewhat similar). *By this, we already implicitly define **the specific laws of relations between properties*** within a certain science or theory.

[2] Infinity, unlimitedness, consciousness, essence, Absolute, and several other categories, which lose the status of sign, i.e., the capability of distinguishing, in their limit form are also attributable to such "limit categories". Some of the paradoxes of this kind are long known, in particular, whether God can create a stone so heavy that He himself cannot lift it.

[3] Diophantine equations are polynomial equations in any number of variables, for which all the coefficients and all the solutions must be integers.

[4] This theorem proves that a formal system **F** (sufficiently complex) cannot concurrently be complete and consistent. Note that our currently available knowledge about Reality in a certain way is reflected in the theoretical concepts.

[5] According to Heisenberg, a quantum object before a measurement must be imaged as a certain potentiality, trend, or possibility, which is quantified as a probability. Note that the probability in his view has a status of a"newkind" of objective physical reality residing in terms of Aristotle's philosophy somewhere "halfway between the massive reality of matter and the intellectual reality of the idea or the image".

[6] The notions "objective" and "object-based" are frequently regarded as synonyms. As is traditionally believed, exclusion of the subject from a scientific paradigm automatically makes science objective (in the sense that it is independent of the subject). However, selection of how an object is described is a subjective event. For example, a ball as a physical body can be described in both classical and quantum representations and these descriptions and, consequently, the objects differ depending on the choice of subject.

[7] A time-limited harmonic is a momentum comprising an infinite number of harmonics.

[8] No doubt that there have been also other observer's reference frames during establishment of physical Reality; however, their distinguishing is what physicists are to do.

[9] In semantic analysis, a creative process always has a single *Subject* unlike several *objects*. The subject may be *detailed to different levels,* which form a hierarchy and are distinguished in a rational manner. The terms "meta-subjects and proto-subjects" (the array of levels to which the Subject is detailed) must be understood in this sense only. In essence, **the Subject is an attribute of the integrity of Reality at any level of its detailing.** The very idea that "subjects" are multiple leads into deadlock. The subject is *indefinable* in any concept of a classical type. A theory of a classical type should study the "subject–object" relationship (*interaction* between the definable objects and *indefinable* subject, which is senseless). However, the principles of integral (subject-based) and analytical (object-based, multiple) understanding of Reality are *complementary* to each other and cannot be *concurrently* implemented. This makes senseless the very statement of psychophysical problem.

[10] Which is evident from the very definition of sign.

[11] The consciousness confined to a particular observer's reference frame.

[12] Genetic, social, or others.

[13] The conditions of attribution can be rather arbitrary for the subject acting as a demiurge relative to the mental map he generates. In particular, one can deny that some races or ethnical or social groups represent a subject; then, certain actions

on these groups are permitted within the class of "objects" whereto they are placed (for example, "a slave is a speaking tool" in the ancient Rome). It is possible to regard animals as subjects: recall how reality is depicted from the "standpoint" of a horse in the *Strider* (Kholstomer) by Lev Tolstoy as well as even inanimate objects can be "awarded" the rank of subjects (*Five Peas from a Pod* by Andersen). In terms of mythology, the forces of nature are typically regarded as subjects (Zeus, Helios, Demeter, Hephaestus, and others as ensoulment of forces of nature).

[14] According to Florensky, the world is one whole and this is a "complex variable", which can be only conditionally partitioned into its real and imaginary portions.

[15] Note that this world itself is regarded as material and the notions, as ideal.

[16] For example, the attitudes incomprehensible to us are in essence objective principles that organize our behavior, which may become objects of consciousness within a new subjective reality emerging as a result of a creative event or "insight".

[17] Here, each coordinate of the vector corresponds to the intensity of a certain property.

[18] Any signs take the sense only when stable meanings are attached to them.

[19] This is also important for adequate understanding of an individual behavior by the other "subjects". If an individual uses rather improbable interpretations in his/her life, he/she looks at least strange.

[20] The absence of the common components in individual mental maps (the "signs" have no common meanings) makes any communications between individuals impossible; thus, they cannot be ascribed to the same mentality.

[21] That pass through the "center of gravity" of the beam.

[22] That is, among the individuals capable of *adequately understanding* one another in their communications. This does not mean that they have *absolutely same views*

on the surrounding reality in the area of our interest. However, they can *adequately understand the positions of the other representatives of the same mentality.* For example, the mentality in marketing characterizes a certain segment of consumers living under similar social conditions.

[23] The sum of squares is taken in order to avoid the sign of projections since we are interested here only in the direction.

[24] A characteristic equation determines the eigenvalues and eigenvectors of linear operators. It is derivable from both the theory of matrices [110] and statistics [111]. Since the dispersion of the estimates given by representatives of a certain mentality forms a dispersion ellipsoid or a quadratic from in the semantic space, it is reducible to a canonical form by solving the same characteristic equation. Thus, we can give both the algebraic (vector)and statistical interpretations to the states of the object. The directions of the vectors that describe the states in a statistical interpretation coincide with the directions of the principal axes of the ellipsoid of dispersion of the object's individual estimates.

[25] In physics, the complementary states of this kind are "motion" and "rest". Determination of the variable, such as velocity, is associated with the former and coordinates, with the latter. It has been known since ancient times that rest and motion are undeterminable simultaneously. The well-known paradoxes of Zeno, an ancient Greek philosopher, demonstrate their logical incompatibility.

[26] Since everything that we have at our disposal is the signals from the "object-based world", our mental representation of it is actually *the map* of this world.

[27] Here, representations of physical objects, psychological and social attitudes, and so on, i.e., everything that is definable in the mental map *via* their *significance,* may act as objects.

[28] More precisely, the probability of a particular selection.

[29] For example, the temperature equalization in a system is explained *via the principle of increase in entropy* and the laws of geometrical optics, *via the principle of least work.*

[30] The projections of all vectors describing objects onto this direction must precisely correspond to their *desirability* for the studied mentality.

[31] Desirability means the *attractiveness* of an object for an individual.

CHAPTER 8

Consciousness and Reality in the Oriental Tradition

Я тку ковер из вечных нитей, вы тот узор не повторите.
Но суть ясна: ищите Путь, и встретимся когда-нибудь.

I weave the carpet of eternal threads; its pattern you cannot repeat
The point is clear—search for the Way and we will sometimes meet.
Midnight thoughts

Abstract: Two types of reality representations are possible-spacetime, describing reality as processes, which may be referred to as "object-based" and corresponds to our unconscious perception, and "subject-based", which is associated with the integral time-independent experience inseparable into objects. The latter forms the foundation for oriental philosophy and is attainable using various psychotechniques. As has emerged, the description of different quantum states in physics requires an analogous approach and this congruence has been mentioned by many outstanding scientists.

Keywords: Consciousness, Nirvana, Quantum mechanics, Reality, Samsara, Space, Sunyata, Time.

INTRODUCTION

Konstantin Tsiolkovsky [112], Vladimir Vernadsky [113], and Aleksandr Chizhevsky [114] in their works searched for the answer to the question of what is the place of Man in the Universe, Man's connection to the endless Cosmos, and the role of this weak creature in the epic performance referred to as the Creation of the World. The very origin of life is the most miraculous event and results from a coincidence of events so unlikely that it looks as someone's caprice. The fact that the man has succeeded in elevating to the level of a reflecting being, attempting to know himself and the surrounding world and to conceive the unity with it, is even more amazing. Einstein, who never accepted quantum mechanics but intuitively believed in the integrity of the world, posed the question of how a

Sergey P. Suprun, Anatoly P. Suprun & Victor F. Petrenko

mouse observing events in space could influence these events. Socrates, a Greek philosopher, asserted, "…know thyself and you are going to know the world", assuming the congruence between the microcosm of a human being and the Universe or, in terms of mathematics, that these infinities comparable in their order are the systems of the same level of complexity. The statement that "the Inner is similar to the Outer; the Small is similar to the Great" is ascribed to Hermes Trismegistus, a mythological protector of medieval alchemists (whose image unites the traits of the Greek god Hermes and Thoth, the ancient Egyptian god of wisdom). The very thought about the likeness of the human microcosm and the macrocosm of the Universe has been a universal idea in many cultures both in the West and in the East. In particular, the Hinduism Vishnusaratantra expressed the idea of similarity between the human existence and the cosmos as "that is Within is equal to that is Beyond". The idea of inner and outer isomorphism can be understood in different ways, for example, *via* regarding the substances of spirit and matter as independent ones (Christian Wolff and René Descartes) in a monistic form, be it idealistic or materialistic. Idealistic monism, originating from Plato, understands the world of objects as "shadows in a cave", which reflect the "world of pure ideas"; in the religious Hinduism and Judaism concept, as "the world of the dream of Shiva and Yahweh"; and in Buddhism, as "Maya", an illusion or projection of human consciousness. First and foremost, works by Hegel, Schelling, and Schopenhauer give a philosophic interpretation of idealistic monism by representing the world as a manifestation of the development of the Absolute Spirit. Currently, a holographic model of the Universe (Bohm [115], Pribram [116], and Talbot and Lipton [117, 118]) suggests similar interpretations; they regard the world as a certain implicit Reality (as a holographic matrix) with yet vague nature. Materialistic monism with the primacy of object-based reality (for example, Marx's economics) regards the human personality as a form (ensemble) of social relationship determined by the economic basis (which as if has created this).

"Stepping down from the level of relationship between the microcosm and macrocosm (the Inner and the Outer) to a more local level of social relationship and psychological processes, we encounter a similar "interiorization" concept (in sociology, Durkheim [119] and psychology Piaget [120] and Wallon [121])" [122].

In the context of the above reasoning, the relationship between the microcosm and macrocosm, as well as human life and cosmogenesis, can be considered in a case study of the phenomenological process of dying or entering the bardo (the "intermediate transitional state" between death and rebirth in Tibetan [123]) to compare this to a hypothetical observation of a person falling into a black hole. In physics, it is believed that the gravitation of such an object leads to the curvature of space that prevents any microparticles or bodies having entered there to come out. A comparison of these two descriptions makes sense because both processes are associated with the disruption of the object-based spacetime model of reality.

Buddhist Interpretation of the Relationship Between Microcosm and Macrocosm

An adequate translation of the content of oriental philosophy and psychology into European languages is rather problematic. This stems from the fact that they are described in the terms and metaphors coined mainly from the 5th century BC to the 10th century AD and formulated under conditions of the specific mentality determined by the culture, language, traditions, and so on. Furthermore, this considerably hinders the collation of these ideas and the corresponding analogous ideas developed currently. The parallels between the attitude to reality in oriental philosophy in ancient times and the current problems in natural sciences, intuitively perceived by many scientists, have been discussed from different perspectives. In particular, Maslow [124], Grof [125], Wilber [126], Tart [127], Frager and Fadiman [128], Kozlov and Maikov [129], and others paid attention to this in their works in the area of transpersonal psychology.

Robert Oppenheimer noted, "…we may have in mind that the general notions about human understanding and community which are illustrated by discoveries in atomic physics are not in the nature of things wholly unfamiliar, wholly unheard of, or new. Even in our own culture they have a history, and in Buddhist and Hindu thought, a more considerable and central place. What we should find is an exemplification, an encouragement, and a refinement of old wisdom" [130];

Niels Bohr asserted that "For a parallel to the lesson of atomic theory regarding the limited applicability of such customary idealizations, we must in fact turn to quite other branches of science, such as psychology, or even to that kind of epistemological problems with which already thinkers like Buddha and Lao Tze have been confronted, when trying to harmonize our position as spectators and actors in the great drama of existence" [131]; and

According to Werner Heisenberg, "…the great scientific contribution in theoretical physics that has come from Japan since the last war may be an indication for a certain relationship between philosophical ideas in the tradition of the Far East and the philosophical substance of quantum theory" [21].

As for the Russian philosophy, these statements are comparable to the views of the representatives of postneoclassical rationalism, in particular, Stepin [103], Lektorskii [132], Kasavin [133], and Mikeshina [134].

It is of interest to compare the worldview that has established in western philosophy to the Buddhist approach, which considers the world from two perspectives, Nirvana and Samsara. In the latter approach, the reality is presented to an ordinary (samsaric) consciousness in a process-dependent object-based spacetime variant. This variant excludes any final cognition at least over any finite time interval. In the case of a "sacral" representation (Nirvana), which is timeless, this state is unrepresentable by any process, including semiosis. In particular, Nagarjuna[1] asserts that language is fundamentally unable to describe the Reality in an apt manner since any tools of language are inadequate to it similar to thinking itself, which operates limited concepts and categories. Strictly speaking, the Buddhist approach consists in that the truth cannot be reached and described in terms of discursive thinking. Consequently, any system of philosophy or any dogmatic Buddhism statement is only the upaya ("skillful means"), *i.e.*, the pointers of the way or the methods helping in this way rather than the result of the way. "Therefore, all things in their fundamental nature are not namable or explicable. They cannot be adequately expressed in any form of language" [135].

The same problems associated with the language in science were described by Heisenberg: "But the problems of language here are really serious. We wish to speak in some way about the structure of the atoms and not only about the 'facts'—the latter being, for instance, the black spots on a photographic plate or the water droplets in a cloud chamber. But we cannot speak about the atoms in ordinary language. Therefore, the correlation between the mathematical symbols of quantum theory and the concepts of ordinary language at this point is unambiguous" [21].

Here, the difficulty lies in that the sign in semiotics acquires any meaning *only in opposition* when it distinguishes and sets the boundary between "something" and "the other" (Terminus is the Roman god who protects boundary markers). Correspondingly, several marginal categories are not the signs in their essence and their use in the discussion is improper and entails logical contradictions. In particular, these categories include "absolute", "infinity", "continuum", "unlimited", "uniform", "reality", "being", "nothing", *etc.*, since they cannot be counterposed to any equal category in terms of semantics. Several paradoxes from the modern theory of sets associated with the term "infinity" or from philosophy associated with the term "Absolute" are adequate examples here. Ascribing any hallmark to the latter inevitably leads to a logical contradiction, for example, "Can God the Almighty create a stone so heavy that He won't be able to lift it?" Russell asked Gottlob Frege,[2] a famous philosopher and mathematician, a similar question related to the theory of sets. As has appeared, many mathematical problems associated with the notions of this kind are indescribable at all with algorithmic languages (in particular, certain problems connected with effective finiteness or infiniteness). A mathematician once joked, "When a dream comes true or infinity is reached, the result not always coincides with what has been expected. There can be everything from "Yes, it finally happened" to "What the hell is it?" [137].

The attempt to remedy this by counterposing marginal categories, such as "Being" and "Nothingness", also fails to resolve this problem. *The existence* of Being (which actually does exist) in this case is counterposed to *the existence* of Nothingness (which actually *does not exist*); however, the very opposition implies the existence. This is an evident logical contradiction: on the one hand, Being is considered as the Absolute, *comprising everything* and, on the other hand, this Being is *counterposed to something represented* by Nothingness, *not contained in the Absolute* (otherwise, this is a part of the Being).

Chan Buddhists believe that the discursive thinking relying on the binary opposition itself gives rise to binary relations *via* separating an integral whole into components and counterposing them to each other. This disrupts the oneness and integrity of being resulting in the alienation of man from his inner nature so that he opposes his individual self to the overall objective reality, which he starts to perceive as the external object-based reality.

Confucianism believes that the sign system in general and the verbal texts in particular are of a paramount importance, while the Chan Buddhists believe that all prescriptions and the texts containing them are false for the particular reason that they are verbalized. This attitude was distinctly formulated by Bodhidharma (in Japan, known as Daruma), the legendary Indian missionary, "Not founded on words and letters" [138]. An analogous approach is evident in Taoism, where Tao (the Way) acts as something unverbalizable and indescribable in terms of differentiating characters, "shapeless", and "boundless". This was further developed in Mahayana Buddhism, all branches of which agreed that the true reality was inexpressible *via* linguistic tools [139]. According to Nagarjuna, the whole prohibits any definitions in the form of notions or words since they can only "dualize" the reality rather than take possession of it. The "Awakening" becomes possible only when an adept becomes free from any attachment to word and sign. That is why the Lankavatara Sutra, which forms the background for both Madhyamaka and Yogachara, refers to the words of Buddha as "wordless". A nonverbal cognition of genuine reality has been tightly associated with the major categories of Buddhist philosophy, namely, nirvana, sunyata (emptiness), and anatman (nonexistence of the self). The paradoxes of cognition are deepened in that we are totally within the fundamentally limited (sign-based) environment. The point here is that both the language (the second signal system according to Pavlov) and sensations (the first signal system) represent *a signal system.*[3]

Here, we can formulate some questions unlikely to be resolved. First, assuming that our sensations are objective, we have "to objectify" the source of these sensations as an "object", a certain factor that unites the totality of properties generating these sensations, thereby mentally constructing an ideal world of the things "beyond" our sensations. However, this constructed world logically inevitably emerges to be transcendent for us since we are completely enclosed in the world of sensations; thus, the so-called psychophysical problem is contradictory at its very setting. In the *Critique of Pure Reason,* Kant spoke about the theoretical and abstract (as he saw it) possibilities of cognizing the things in themselves (reality as it is) and noted that it would be necessary for this purpose to get rid of the sense experiences (space and time) immanent in the subject and the mental categories and then, to acquire another kind of non-sense experience (intuition). Schelling agreed with this way of cognition and referred to it as "intellectual intuition" in line with tradition. Schopenhauer mocked this Schelling's wording, yet admitted a mystical understanding of the "thing in itself" (according to his doctrine of will). Henri Bergson (in *The Creative Evolution*) considers an intuitive channel of information allowing one living being to sense the state and morphology of the vital organs of another living being *via* experiencing these states and as if simulating these states *via* the own psyche. This resonates with the fact that Hinduism and Buddhism admit the existence of nonmediated cognition along with the five senses achievable *via* meditation and "nondual vision" that removes the opposition of subject and object [141]. However, this way of cognition is beyond the frame of the classical object-based approach since it excludes the very cognizing subject from its paradigm.

Another difficulty in the comprehension of Reality is a processual *spacetime representation* of the mentally constructed object-based world. As for the independent existence of space and time, as early as Leibniz wrote: "For my part, I have said several times that I hold space to be something merely relative, as time is, taking space to be an order of coexistences, as time is an order of successions. For space indicates an order of things existing at the same time, considered just as existing together, without bringing in any details about what they are like. When we see a number of things together, one becomes aware of this order among them. As for those who imagine that space is a substance, or at least that it is something absolute, I have many demonstrations to show them to be wrong" [142].

In the 20th century, the position of Poincare, one of the founders of the relativity theory, was no less radical. In his view, space and time are one of the possible representations of reality accepted, thanks to a tacit agreement as the simplest under certain conditions and the most convenient projection of our sensations. He noted, in particular, "Is the position tenable, that certain phenomena, possible in Euclidean space, would be impossible in non-Euclidean space, so that experience, in establishing these phenomena, would directly contradict the non-Euclidean hypothesis? For my part I think no such question can be put" [19]. He also adds another comparison to this one, namely, "Then what are we to think of that question: Is the Euclidean geometry true? It has no meaning. As well as ask whether the metric system is true and the old measures false? whether Cartesian coordinates are true and polar coordinates false? One geometry can not be more true than another; it can only be more convenient" [19]. What is Time? Time is the past, present, and future. The present exists only with regard to the past and future and the past and future, in turn, exist only relative to the present (*i.e.*, everything *is related to* something other, which is an axiom of semiotics and, according to Nagarjuna, has no own existence). However, the past has already gone and the future has not still come into existence. Hence, Being of the present is defined by nonentities, making it fictitious and empty as well.

Does time flow? Does it have a direction, a beginning, or an end? Sir Arthur Eddington introduced the phrase "the arrow of time" and this is one of the greatest mysteries. We believe that a certain succession of events is natural, for example, people become old, glasses break, and the candle burns down. This gives us the feeling that the events are irreversible in time but fail to give any understanding of the underlying causes. Time in the laws of physics (which as if describe reality in an adequate manner) is symmetrical; moreover, it is a scalar. This means that there is no difference between the past, present, and future. As for the quantum electrodynamic equations describing the evolution of elementary particles, the ($\pm$) sign may be related to time; correspondingly, they "come" from both the future and past.[4]

According to Schilpp [143], "… all that each of us perceives as the past, the present, and the future appears merged in spacetime. Each observer, moving with the flow of the own time, faces a kind of "layers" of spacetime and sees in these layers the representations of the material world that replace one another although the indivisible wholeness of all phenomena comprising spacetime foreruns his (observer's) knowledge about them". Or, as Kennett [144] put it, "… many people believe that time passes but it actually remains where it was. This idea of "passage" may be regarded as "time" but it is a false representation for if you see it only as "passage", you cannot understand that it remains where it was".

Einstein's general theory of relativity (GTR) describes the Universum in spacetime as a whole object, or "givenness". However, everything is givenness for God. An attempt of describing the Universum "from outside" as an object brought about a great set of problems, such as the so-called cosmological paradox.

It follows from the GTR (according to the Doppler shift in the infrared spectrum range) that our Universe is not stationary and cosmology puts the age of the Universe at about 15 billion years. The Big Bang, which gave birth to the Universe, belongs to the category of events; however, this category is out of our formulations of the laws of nature. This situation is similar to a description of a quantum system, with waves that have neither beginning nor end. It is high time here to recall St. Augustine: "For whence could innumerable ages pass by, which Thou madest not, Thou the Author and Creator of all ages? or what times should there be, which were not made by Thee? or how should they pass by, if they never were? Seeing then Thou art the Creator of all times, if any time was before Thou madest heaven and earth, why say they that Thou didst forego working? For that very time didst Thou make, nor could times pass by, before Thou madest those times" [24]. The medieval question of what God had done before creating the world again became relevant, this time for physicists. The emergence of Being from Nothing is a creative event. However, creativity runs into conflict with time in terms of physics. Once we are able to describe the alteration in the form in finite theories (according to Hilbert) with limited and constant content, how then is it possible to describe the changes in the content itself? A classical theory with completely undefined or spontaneously changing axiomatics cannot be created. It was as early as Plato when the reason and truth were linked to the access to Being, the unaltered reality underlying the Development (coming into being). However, Plato, recognizing a contradictory character of this position, came to the conclusion in his *Sophist* that both Being and Development were necessary. The ancient atomists encountered the same problem: to admit the emergence of new content, Titus Lucretius Carus had to invent the notion of clinamen, an unpredictable swerve of atoms, *i.e.*, a sort of indeterminacy principle:

"We wish thee also well aware of this:

The atoms, as their own weight bears them down

Plumb through the void, at scarce determined times,

In scarce determined places, from their course

Decline a little—call it, so to speak."

[Lucretius. *On the Nature of Things* (Translated by Leonard, W.E.), 2017.]

Interestingly, this indeterminacy as an integral part of reality was mentioned by Einstein in his work on the emission and absorption of radiation by excited atoms [145]. There, he asserted, "...time and direction of elementary processes are defined in a random way". This is unavoidable since creativity (as the opposition to determinacy) is beyond the science of physics, being associated with physicists rather than their theories. The arrow of time is incomparable to evolution as a process of changes in the content, whereas creativity belongs to the subject rather than the object-based world. Perhaps, that is why researchers capitalize the word Nature in this context, implicitly regarding it as a subject. Ludwig Boltzmann, an Austrian physicist, introduced the physical notion of time as being connected with evolution, which was his lifelong mission. Neither clinamen nor creativity was considered by academic science because they violated the principles of locality and causality in this world. However, the experiments by Aspect *et al* . [146] associated with the so-called EPR paradox, *i.e.*, quantum teleportation of states [30], obviously show that Mother Nature does not prohibit such processes.

Another possibility to define the arrow of time is provided by the so-called reduction of the wave function as a physical process that takes place at the very moment when an object is perceived. This fact actually brought the overall physics to the brink of the debate. In his *Philosophical Problems of Quantum Physics,* he writes that it is necessary to reject the ideas about an objective timescale common for all observers, as well as about the events in space and time that would be independent of our capability of observing them. Heisenberg emphasizes that the laws of nature describe *our knowledge* about elementary particles rather than the particles themselves, that is, actually, the content of our mind.

In 1958, Schrödinger published a short book titled *Mind and Matter* [147], in which he turned from the new physical results to a mystic view of the Universe complying with the "perennial philosophy" by Aldous Huxley. Among the theoreticians in quantum physics, Schrödinger was the first to sympathize with the Upanishads and the ideas of the oriental philosophy. The views of many modern physicists are summarized by Wigner, a Nobel Prize winner, in his *Symmetries and Reflections: Scientific Essays* [148]; he also relied on the ideas of the Universe as an all-pervading Consciousness. First, Wigner notes that most physicists came back to the acceptance of the fact that the thought (or mind) is a primary substance. Then he asserts that consistent laws of quantum mechanics cannot be formulated without involving consciousness in them and concludes that it is really amazing that the scientific research into the world has shown that the content of our consciousness is the primary reality. A mental experiment—the paradox of Wigner's friend—is also known among physicists; it puts the question at which particular moment the result of measurement of a quantum system is recorded. It is considered an extension of the famous mental experiment of Schrödinger's cat.

Wigner devised this experiment in order to focus on the necessity of involving consciousness in the event of a quantum mechanical measurement. Proietti *et al* . [149] report the experimental results on testing the following three assumptions:

1. Do the observed facts in reality actually reconcile on a global scale beyond spacetime frames?
2. Or the locality principle, which is fundamental in an object-based simulation of reality, is met?
3. And in this case, does the observer have a free choice in the sense that the events are independent, which makes it possible to utilize the classical probability theory in the description of experimental results in the area of quantum physics?

As has been expected the first assumption matched best the obtained results and, in addition (which is omitted by the authors of this paper), the wave function has emerged to reflect the standpoint of a particular observer, which corresponds to the definition of observer's reference frame.

As we see, the partition of Reality into physical and mental counterparts brings about not only logical, but also purely experimental paradoxes, interfering with the understanding of experimental facts. This meets the ultimate psychologism of Buddhism, which does not consider the world as itself but rather as the psychocosm, *i.e.*, the world experienced by *a living being* as an aspect of its psychic experience, as unfreedom and, consequently, as *suffering.* Actually, various worlds have been analyzed by Buddhism as the deployment levels of the consciousness of living beings [150]. A similar idea is also present in the current thought. In particular, a profound book by Rubinshtein, *Man and the World* [151], published half a century after the death of the author, is permeated with the same idea that the world is a form of man's being and the world as suffering. As Einstein put this, "A human being is a part of the whole, called by us "Universe"; a part limited in time and space. He experiences himself. His thoughts and feelings as something separated from the rest—a kind of optical delusion of his consciousness. This delusion is a kind of prison for us, restricting us to our personal desires and to affection for a few persons nearest us. Our task must be to free ourselves from this prison by widening our circle of compassion to embrace all living creatures and the whole of nature in its beauty" [152].

Jacob von Uexküll, a famous biologist and philosopher, calls attention to how diverse can be the world in the consciousness of different living creatures: the pine for a forester is a tree and construction material is the home and shelter for a fox that has a burrow under the pine roots and a plenty of food for a bark beetle. Modern psychosemantics examines these sides of the worldview [153].

Considering again the rules of signal-based representation of reality in our consciousness, note that this representation is its basis, and a spacetime form of information representation. The term "signal" is commonly recognized for characterization of the method of data representation when these data are regarded as the result of measurements (perceptions) of a studied object as a sequence of scalar values (analogous, numerical, graphic, and so on) depending on the changes in the values of some variables (time, spatial coordinates, energy, temperature, and others). The material form of signal carriers (mechanical, electric, magnetic, acoustic, optical, or some other) as well as the form of their representation in physical parameters or processes, is of no importance. In a general case, the signals are described by a functional dependence of a certain information parameter of the signal on an independent variable (argument), such as $s(x)$ or $y(t)$. This form of signal description and its graphic representation are referred to as a *dynamic* representation (*i.e.*, describing the behavioral change in the signal with changes in the argument).

Along with a usual dynamic representation of signals and functions in the form of dependences on certain arguments, a mathematical description of signals with the help of the arguments inverse to those in a dynamic representation is widely used in analysis and data processing. The description of this type is possible because any signal arbitrarily complex in its shape that has no discontinuity of the second kind (lacking infinite values in the interval where it is defined) can be represented as a sum of simpler signals, for example, as the sum of simplest harmonic fluctuations. This description requires the Fourier transform technique.[5] An appropriate example here is the pair "time–frequency".

More strictly, the following operation is referred to as the Fourier transform of function $f(x)$ (in our case, of a signal depending on time):

$$F(\omega) = \int_{-\infty}^{\infty} f(t)e^{-j\omega t}\,dt \ .$$

Thus, the function of time is transformed into the function of frequency; this is a decomposition of the function into its components for different frequencies (here, ω is frequency; j, unit imaginary number; and e, base of natural logarithm). An inverse Fourier transform is put down as:

$$f(t) = \int_{-\infty}^{\infty} F(\omega)e^{j\omega t}\,d\omega$$

In abbreviated form, these relations are described using the symbols of Fourier

transform operator F and F^{-1}: $f(t) \xrightarrow{F} F(\omega)$ and $F(\omega) \xrightarrow{F^{-1}} f(x)$.
Function $F(\omega)$ is referred to as the Fourier transform of function $f(t)$ and, in turn, function $f(t)$ is named an inverse Fourier transform of function $F(\omega)$, or pre-transform.

Accordingly, the functions of amplitude and initial oscillation phases with respect to a continuous or a discrete argument (the frequency of changes in the function at some intervals of arguments in their dynamic representation) describe the mathematical decomposition of a signal into harmonic elements. The totality of the harmonic oscillation frequencies is referred to as the signal's general amplitude spectrum and the set of initial phases, as the phase spectrum. Together, both spectra constitute the signal's full frequency spectra, which give an unambiguous and complete representation of a spacetime form of the signal in a Hilbert space.

Actually, the Hilbert space describes a state implemented as a process in spacetime. Thus, reality appears in our consciousness in an object-based spacetime form as a sequence of sensations and perceptions with the involvement of memory and thinking. In this process, the mental domain (the source of something in our consciousness) is referred to as the unconscious. The transition from the unconscious to consciousness is assumed connected with a change *in the way of reality representation.* Two different ways of reality representation, being fundamentally incompatible, must yet be equivalent in the content they represent. Moreover, they are complementary since the way of reality representation in the unconscious fundamentally cannot be compatible with the representation in consciousness. Indeed, two types of representations are available and can be analyzed—a spacetime representation, where the content is successively implemented, and a holistic representation, where the content is implemented simultaneously (see, for example, *Thinking and Speech* by Vygotsky [154], considering the translation of thinking process into speech). The first representation can be arbitrarily regarded as an object-based one and the second, as subject-based one since the latter shows an integral time-independent state of the subject. The absence of continuous time in a Hilbert space does not exclude that the states can be indexed and ordered as a certain sequence. Poincare in his *Sur la théorie des quanta* (On Quantum Theory) [153] assumes that the assertion of a discrete behavior of the set of possible states of any isolated physical system is pertinent to the Universe as well, putting down that the Universe, thus, must jump from one state to another but still remains unchanged between these leaps, and the different moments when it retains its state would be indistinguishable from each other; correspondingly, this leads us to discrete time flow, that is, the atoms of time [155]. Note that the time in the space of states is of an evolutionary nature and discrete *versus* the physical time, which is "continuous". The former specifies the order of changes in states (as well as the contents of the developing processes) and the latter, defines their duration.

Thus, both the "experiencing" of Reality in the Hilbert space of states and the possibilities of their verbalization radically change:

1. The unusual experience of the integrity of the world emerges since both an object-based decomposition and classical spacetime representation become unfeasible. Consequently, any ego experiences and ego itself are absent;
2. Physical space and time and the existence of nonlocality are absent, including the local sensations. The absence of rigidity (inertness) is experienced as instantaneousness in any changes both physical and mental, including the verbal processes, which is experienced as the inability to express in words an immediate experience of reality; and
3. The intuitive grasp of the world is experienced as a *state* resulting from the identity with it rather than any logical thought or perception of the state in evolving spatiotemporal form.

Now compare the perception of *"the space of states"* to the experiences that appear in a meditation (Dhyana) and correspond to the state that in the oriental tradition (Buddhism, Taoism, and Yoga) is referred to as Satori, Zen, Awakening, Nirvana, Samadhi, and so forth [156].

1. The Chan principle says, "Do not rely on words and letters" and "Be free from name and form". It is asserted that since the words (and the notions expressed with words) are unable to adequately reflect the true reality, the insight into this reality requires the return to the whole undivided source of experience, residing in the deep layers of the mind unaffected by verbalization. That is why all trains of thought that are not conductive (in one or other way) to attaining such psychological experience are regarded as useless, unnecessary, and even adverse. It is also believed that it does not actually matter how to name this genuine reality, be it alaya-vijnana (in Vijnanavada) or sunyata (in Madhyamika), while overcoming the emotionally and psychologically "shadowed" state. On the contrary, the question on how it is possible to implement this state of consciousness and what are the ways to identify oneself with the genuine reality becomes much more important and essential [157].
2. The true reality acts as something certain that cannot be verbalized and

described in terms of differentiated characteristics, as something "mysterious", "miraculous", "incognizable", "shapeless", "boundless", and so on [157]. That is why any cognitive approaches to its comprehension are excluded, whereas the major methods are psychological practices, as a rule, different types of meditation. In Chan Buddhism, meditation is performed *via* focusing the consciousness, freed from any images, at one point (ekagra in Sanskrit and yi-nan-xin in Chinese) in combination with the maximum relaxation and stabilization of consciousness, removal of psychic stress, and attainment of the most balanced state. The meditation typically commences from a purposeful concentration of attention at one point and intensive "look" with the mind's eye into "emptiness" trying "to devastate" the consciousness to a complete absence of any thoughts or images. This state is referred to as "one-pointed" consciousness (yi-nan-xin), "the consciousness without thoughts" (wu-na--xin), or "non-consciousness" (yi-xin) [157].

3. Note that the very meditation as a tool cannot be the cause of Satori, being beyond the limits of Samsara[6] (beyond the boundary of individual consciousness) but rather increases the probability for *the type of consciousness* to spontaneously switch. Evidently, any attachment to locality in any form (embodiment, ego, instinctive sphere of individual, personality, and so forth) blocks this possibility. The Awakening (Satori) is implemented as a spontaneous creative event with an uttermost pronouncedness, as a jump from a limited space to infinity, and is described by Zen Buddhists with the metaphor "the barrel head suddenly falls off".

Chuang Tzu gave the following description for this experience: "I smash up my limbs and body, drive out perception and intellect, cast off form, do away with understanding, and make myself identical with the Great Thoroughfare. ... Gazing, we do not see it; we call it dim. Listening, we do not hear it; we call it inaudible. Groping, we do not grasp it; we call it subtle; these three [properties] do not allow ultimate scrutiny, for indeed, merging, they become one. Its rising is not bright, nor its setting dark. Branching out in shoots innumerable that cannot be defined, it returns again to nothingness. This may be called giving shape to shapeless, forming an image out of nothingness; this way may be called a vague likeness. We meet it, but do not see its front; we follow it, but do not see its back. If by seizing the Way of antiquity, we direct the existence of to-day, we may know the primordial beginning. This may be called: [unraveling] the clue of the Way" [158].

Note that the Samsara (object-based world) and Nirvana in Mahayana Buddhism are in essence identical since they are two types of reality representation rather than two realities. Unlike an instant experience in Nirvana, the cognition in Samsara is an infinite process allowing for sequential approaching the truth.

According to the Yogachara school, "In our mind the uncountable things caused by differentiation originate... People perceive these things as an external world... That seems external, doesn't exist actually; what we see plurality in, is actually just our mind: the body, property and all mentioned above—all that is only the mind and nothing but the mind, I tell you» [159].

As is evident, the specific features of "experiencing" reality in the space of states and Nirvana completely coincide, allowing for a serious attitude towards the practices and psychotechniques that were developed some 2000 years ago in the frame of the mentioned traditions rather than for regarding them as artifacts of mystic or ecstatic states of consciousness or "extension" of everyday consciousness (or, as a variant, "an altered state of consciousness"). *This is a fundamentally different form and type of consciousness incompatible with the everyday consciousness and in no way representing its modification.* It is quite evident that the psychological techniques assisting such switch from an object-based representation to a subject-based one do not "extend" everyday consciousness (as the relevant literature on transpersonal psychology uses to state) but, on the contrary, narrow the consciousness to one point to leave it. One may ask the question on which particular representation of reality is actually primary and which is its derivative. We have discussed above the paradoxes that result from an object-based representation. As for the subject-based representation, these problems do not simply appear just because objects, space, and time are absent. Is it feasible to reveal hidden subject-based roots of the genuine reality in an object-based representation? As it has emerged, this is possible.

Consider a pulse with a duration T (from 0 to T) with the spectrum described by a Fourier integral. At a frequency ω, it contains a harmonic component defined over *the entire infinite time axis* (overall past and future). As can be shown, the sum of all these components for any time moment $t < 0$ is zero *except for $t > T$*. Actually, this implies that the harmonic component *had existed before this pulse emerged,* that is, this is an evident *violation of the law of causality*. Thus, the transition from an *integral* (time independent) description of Reality to a spacetime representation *inevitably brings about the violations of causality and locality.* This is confirmed by experiments on quantum teleportation, in which two objects within the same state (physicists refer to them as "entangled") are connected in a nonlocal instant manner.

A short time Fourier transform is usually used for a time–frequency localization of structural elements:

$$S(\omega,b) = \int\limits_{-\infty}^{+\infty} s(t)e^{-i\omega t}w(t-b)dt,$$

where $S(\omega, b)$ is the Fourier transform of a signal $s(t)$ multiplied by the window (local) function $w(t - b)$. Using this function, we actually "screen" the signal in a certain time window, which naturally corresponds to a certain spectral window.

This significantly increases the potential of this method as compared with the classical Fourier transform but is not free from certain shortcomings. According to the principle equivalent to the Heisenberg uncertainty relation, by using this transformation it is impossible to state that the signal contains the frequency ω_0 at

time moment t_0; however, we can assert that the interval (t_1, t_2) contains the frequency band (ω_1, ω_2).

Thus, the spectral analysis makes it possible to determine only the time intervals during which the spectrum contains frequency bands. This problem is referred to as an Entsheidungsproblem (decision problem) and is associated with the width of the used window function: a narrow window gives better time resolution and a wider one, better frequency resolution. Correspondingly, adjustment of the window width and shape (which is analogous to the function of attention) may be necessary for the analysis of the entire signal.

This allows for understanding how *the experience of time* emerges in "limited" consciousness, being, conceivably, absent there. Instead of considering the spectrum of *the entire signal* (*i.e.*, over infinite "time interval"), we confine ourselves to the spectral decomposition of this signal into *sequences of finite intervals*. The main point of this representation of the time dynamics of a phenomenon is that *the Fourier transform has not been completed* since a full and accurate representation of any signal demands that it is considered *over the entire infinite time axis*. We actually "chop" signal $S(t)$ into fragments in the time domain or (which is equivalent) through a spectral window in a frequency representation, *i.e.*, we *filter* the signal.

However, only some part of the spectrum of the "cut-off" pulse passes through a narrow-band filter; *the changes in amplitudes and phase shifts* of the components are determined by the filter. However, the higher the **quality factor** of the circuit, the more **inertia** it has; correspondingly, the longer time interval is needed for *any change in the amplitude* of oscillation in the circuit. Thus, a decrease in the Δt of time quantum and the size of filter cells allows us to increase the error in the estimation of the signal's amplitude and energy. This is equivalent to the Heisenberg principle of uncertainty in quantum physics,[7] *i.e.*, it is our perception that gives birth to the rigidity of the processes.

Then it is clear that it is not an independent property but one of its characteristics along with its intensity and quality. This property characterizes the cut-off parameter, Δt, of the event of perception *via* the Heisenberg uncertainty principle because it determines *what we actually perceive.*

The paradoxes of quantum mechanics obviously suggest that it is inadequate to ascribe an object-based spacetime form to Reality. According to Nagarjuna, none of the elements of samsara has its own being since this being has been borrowed from some other elements, which, in turn, borrow it; this is by no means a true

being similar to that the borrowed money is by no means a fortune. Correspondingly, all elements and the structures formed of them are entity-less, or empty [158].

"What a whiff of horror comes from this insistent and adamant denial of everything, even the most highly esteemed and carefully preserved ideas of Hinayanists. 'What should we do?' exclaims Aryadeva, the most brilliant of the exegetes of this doctrine, 'Nothing exists at all! Even the name of the teaching evokes fear'..." [159]. However, the paradoxes disappear once we assume that the states of reality (naturally, limited by our, "observer's, reference frame") are directly presented to our unconscious and these states in a spacetime form are translated to our consciousness. According to Fedor Shcherbatskoi [159], in the view of Nagarjuna, "there is not a shadow of difference between the absolute and phenomenal, between nirvana and samsara. The Universe considered as a whole is the Absolute, while the universe considered as a process is a phenomenon."

Shcherbatskoi also notes interesting parallelism between some views of Hegel and Nagarjuna: "Hegel in *The Phenomenology of Spirit* doubts the ability of common sense to evaluate an object known to us from experience; all that we actually know about this object is its "thisness" (being a given this), whereas its entire remaining content will be just an attitude. This is actually the precise meaning of the term *tathata* or "suchness" of the Mahayanists, while relativity ... will be the precise meaning for *sunjata.* Then we find there a full application of this method, which reckons that we can correctly determine an object only having obtained accurate data on the other objects to which it is contradistinguished; that the object without this opposition becomes "void" and deprived of any content; and that both opposites are united in a certain highest oneness comprising both of them" [159]. Modern physicists are no strangers to similar reasoning: "An elementary particle is not an entity that exists independently, but rather it is a set of relationships that reach out to other things" [160]. "The world thus appears as a complicated tissue of events, in which connections of different kinds alternate or overlap or combine and thereby determine the texture of the whole" [21].

Currently, the experiments on quantum teleportation of states are widely debated in physics. It is of interest to compare these experiments to the techniques for the transfer of consciousness in Tantric Buddhism (one of the highest Yogas of

Naropa [161]). This technique is used in the Tibetan tradition of Dzogchen Buddhism[8] to reach Nirvana and is practiced in the so-called bardo state (an intermediate, transitional, or liminal state between death and rebirth [123]). It is interesting to compare the experiences of the patients after clinical death and the descriptions of after-death experiences from the *Bardo Thodol*[9] Tibetan book, on the one hand, and some physical representation of cosmological phenomena, on the other hand, in particular, the experiences of "falling into an abyss", moving in a "tunnel" or "tube", or a flash of bright light, the so-called Rigpa[10] or Dharmata[11] [162, 163].

Find below some details related to this issue.

In order to visualize object-based mental representations, the so-called mental maps of consciousness are constructed by psychologists; these maps present Reality to the subject in an object-based spacetime form. Since the models of reality in physics are constructed in the very form, this reality is represented in our consciousness, these models actually implement a physical component of these maps. It can be shown based on purely psychophysical patterns that the spaces of these maps possess a Minkowski metric [51, 164], the very same metric obtained in Einstein's *special theory of relativity*. Taking into account the rigidity (inertness and stability, mentioned above) of sensations, these spaces thus correspond to pseudo-Riemannian spaces of the GTR, which, in particular, considers the "objects" as black holes. Thus, the similarity of these descriptions is not merely accidental but rather stems from the common roots because the only way out from the mental map having the same metric as Einstein's spaces is the analogs of black holes. As is known, the black hole is a region of spacetime with a gravitational effect so strong that no objects, including those moving with the velocity of light, are able to escape from inside of this hole. The boundary of this region is referred to as the event horizon and its characteristic size (depending on its mass), is the gravitational radius [165]. This radius for the simplest variant—the so-called spherically symmetric black hole—amounts to the Schwarzschild radius,

$$r_s = \frac{2GM}{C^2}$$, where C is the velocity of light; M, body mass; and G, gravitational constant.

Some exact solutions for Einstein's equations suggest a theoretical possibility that such spacetime regions exist; the first of them was obtained by Karl Schwarzschild as early as 1915. The two most important features characteristic of static black holes are the event horizon (possessed by any black hole by definition) and singularity,[12] separated from the remaining Universe by this horizon. Unlike the rotating hole, the existence of a static black hole is most improbable; correspondingly, we consider a charged black hole,[13] which is frequently utilized as a model for rotating holes because their geometries are rather similar. This is also explainable by an extremely complex solution for the rotating hole. This solution was first most briefly described by Kerr in 1963 [166] with a detailed version published by Kerr and Schild a year later. Debney, Kerr, and Schild in their well-known paper [167] comprehensively described the decision-making by Kerr and Kerr–Newman. The presence of a "wormhole", which can in principle connect infinitely many "independent" spaces, holes, be it some "other" universes or distant parts of our Universe, in the event horizon of a charged hole is the main feature that distinguishes between a charged and uncharged non-rotating black holes. We do not consider here the issues of wormhole stability and the role of the so-called "exotic" kinds of matter since these issues are still vague.[14]

It is important to us that the only currently known exit from this object-based spacetime reality is black holes (this is, correspondingly, true for the exit from *a mental map of consciousness*). Thus, a comparison of the states described in the Tibetan book Bardo Thodol (a kind of a guidebook on the after-death stages of "leaving" a particular reality) to the "travel" to a black hole looks quite intriguing (Fig. **8.1**). The parallels emerging in this case may be determined by the process of "breakdown" of a particular mental model of reality at the moment of death. The logical constructs in terms of the dharma theory have been studied by the adepts of Tantric Buddhism, "experiments" included (the chöd ritual, a practice of experiencing a sort of clinical death).

Many physicists (Bohr, Schrödinger, Bohm, and others) have pointed to that the oriental philosophies better match the new physics as compared with the European ones. Grounding their doctrines on the integrity of the world, subject-based nature, and development, the oriental philosophies have succeeded in reaching the most impressive conclusions, frequently with a shade of mysticism. Here, we want to formally compare some physical phenomena of modern physics to their Buddhist analogs. Without this mystic veil, perhaps, the similarity has deeper logical grounds. Anyway, the phenomena associated with the development of integral open systems, be it the Universe or Man, must have many things in common. At the moment, we know how to describe closed systems allowing for the formulation of conservation laws and, correspondingly, for the prediction of further formal changes. The open developing systems, which change their content, must be described in a different manner. Presumably, the oriental experience, logic, and philosophy would give us some tips. Tantric Buddhism has special techniques for attaining the states of this kind aiming to thoroughly examine them and further use them in order to switch to another type of consciousnesses

(Nirvana) *via* transferring consciousness at the moment of death (the so-called phowa[15]).

The tantras of Dzogchen distinguish four bardo states, namely,

(i) The "natural" bardo of this life;

(ii) The "painful" bardo of dying;

(iii) The "luminous" bardo of dharmata; and

(iv) The "karmic" (casual) bardo of becoming.

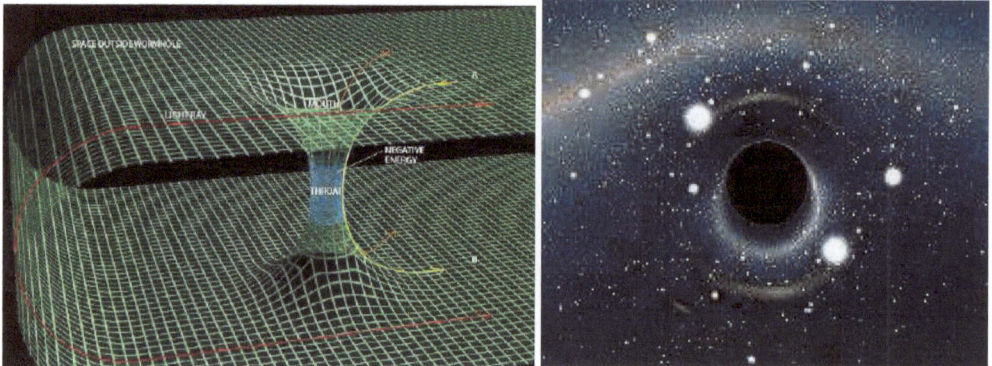

Fig. (8.1). A wormhole and a black hole.

Dzogchen asserts that our ego when dying, first, loses its physical features, which are most rigid, thereby considerably promoting the possibility of switching between the types of consciousness ("recognition of the nature of mind") for a trained adept. Sogyal Rinpoche wrote in *The Tibetan Book of Living and Dying* [163]: "Think of the moment of death as a strange border zone of the mind, a no-man's land in which, on one hand, if we do not understand the illusory nature of our body, we might suffer vast emotional trauma as we lose it, and on the other, we are presented with the possibility of limitless freedom, a freedom that springs precisely from the absence of that very same body. When we are at last freed from the body that has defined and dominated our understanding of ourselves for so long, the karmic vision of one life is completely exhausted, but any karma that might be created in the future has not yet begun to crystallize. So what happens in death is that there is a "gap" or space that is fertile with vast possibility; it is a moment of tremendous, pregnant power where the only thing that matters, or could matter, is how exactly our mind is. Stripped of a physical body, the mind stands naked, revealed startlingly for what it has always been: the architect of our reality. So if, at the moment of death, we have already a stable realization of the nature of mind, in one instant we can purify all our karma. And if we continue that stable recognition, we will actually be able to end our karma altogether, by entering the expanse of the primordial purity of the nature of mind, and attaining liberation" [163].

The third phase of dying (the most important to us) is the passing through the stage of blackness, "like a sky shrouded in darkness". "The arising of the Ground Luminosity is like the clarity in the empty sky just before dawn. Now gradually the sun of Dharmata begins to rise in all its splendor, illuminating the contours of the land in all directions. The natural radiance of Rigpa manifests spontaneously and blazes out as energy and light" [163]. If this moment is missed, the fourth stage ("karmic bardo" of becoming) commences; this is an intermediate state lasting to the next birth. Note that according to Dzogchen doctrine, we first see our past and live through it again at the moment when Rigpa appears and then our future in our "new birth".

Find below the descriptions of the near-death experiences of the patients after clinical death from the *Reflections on Life after Life* by Prof. Raymond A. Moody [168] to compare these descriptions to the Bardo stages.

This near-death experience consists of the following stages independent of gender, age, culture, education, and religion, cited from [168]:

(1) **Ineffability.** All people who have had a near-death experience tell that it is completely ineffable. See a fragment of an interview with such a person: "Now, there is a real problem for me as I'm trying to tell you this, because all the words I know are three-dimensional. As I was going through this, I kept thinking, 'Well, when I was taking geometry, they always told me there were only three dimensions, and I always just accepted that. But they were wrong. There are more.' And, of course, our world—the one we're living in now is three-dimensional, but the next one definitely isn't. And that's why it's so hard to tell you this. I have to describe it to you in words that are three-dimensional. That's as close as I can get to it, but it's not really adequate. I can't really give you a complete picture" [168].

(2) **Dark tunnel.** Often, people feel as if they are moving at a very high speed across a space and hear some sound effect. "I had a very bad allergic reaction to a local anesthetic, and I just quit breathing—I had a respiratory arrest. The first thing that happened—it was real quick—was that I went through this dark, black vacuum at super speed. You could compare it to a tunnel, I guess. I felt like I was riding on a roller coaster train at an amusement park, going through this tunnel at a tremendous speed" [168].

(3) **Out-of-body sensations.** Many people cannot imagine that they exist in a certain state outside their physical body. "As I did, I quit breathing and my heart stopped beating. Just then, I heard the nurses shout, "Code pink! Code pink!" As they were saying this, I could feel myself moving out of my body. ... Then, I started rising upward, slowly. On my way up, I saw more nurses come running into the room—there must have been a dozen of them. ...I watched them reviving me from up there! My body was lying down there stretched out on the bed, in plain view, and they were all standing around it. I heard one nurse say, "Oh, my God! She's gone!", while another one leaned down to give me mouth-to-mouth resuscitation" [168]. Another informant told that "during his experience he felt as though he were 'able to see everything around me—including my whole body as it lay on the bed without occupying any space,' that is as if he were a point of consciousness" [168]. It is very easy to travel in such a state since physical objects present no obstacle and moving from one place to another is almost instant. The respondents noticed that time as though did not exist at all when they were out of their physical bodies. One man told, "Things that are not possible now, are then. You mind is so clear. It is so nice. My mind just took everything down and worked everything out for me the first time without having to go through it more than once. After a while everything I was experiencing got to where it meant something to me in some way" [168].

(4) **The being of light.** Moody writes that the most incredible yet constantly present phenomenon in all cases he studied was the observation of a very bright light. This light seemed initially rather dim and then became ever brighter to eventually turn extraordinary bright. However, this light was not painful to the eyes even then. Perhaps, this is explainable by that the patients at that time lacked any "physical eyes" to make them blind.

(5) **The images of previous life.** They can be regarded as reminiscences yet with certain features that distinguish them from common recollections. First, this is an extraordinary velocity of the change in the chronological sequences of pictures over some moments. "After all this banging and going through this long, dark place, all of my childhood thoughts, my whole entire life was there at the end of this tunnel, just flashing in front of me" [168].

(6) **Border or limit.** "Some patients told that during their moribund experience they approached something that can be regarded as a border or a kind of a limit. Sometimes this looked like a gray fog" [168].

(7) **The vision of complete knowledge.** Moody notes, "Several people have told me that during their encounters with "death" they got brief glimpses of an entirely separate realm of existence in which all knowledge—whether of past, present, or

future—seemed to co-exist in a sort of timeless state. Alternately, this has been described as a moment of enlightenment in which the subject seemed to have complete knowledge. In trying to talk about this aspect of their experience, all have commented that this experience was ultimately inexpressible. Also, all agree that this feeling of complete knowledge did not persist after their return; that they did not bring back any sort of omniscience" [168].

In conclusion, Moody notes that "despite their own certainty of the reality and importance of what has happened to them, they realize that our contemporary society is just not the sort of environment in which reports of this nature would be received with sympathy and understanding. Indeed, many have remarked that they realized from the very beginning that others would think they were mentally unstable if they were to relate their experiences" [168].

Comparing these descriptions to the hypothetical phenomena observable when falling into a black hole, first, let us consider the Penrose diagram for a rotating (or charged) black hole (Fig. **8.2**). The Kerr singularity there has a shape of ring; in a spacetime diagram, it will be parallel to the time axis in the Penrose diagram, *i.e.*, it is time-like. Correspondingly, there exists the possibility that once entering a black hole, we will be able to pass the central singularity escaping its tremendous gravitational forces since only the bodies in the ring singularity moving in the hole's equatorial plane collapse.

As is evident from the full Penrose diagram, the hole of this type has two event horizons, outer and inner, the latter being located closer to the singularity.

The body that is under the outer horizon cannot go outside anymore because of the properties of spacetime there and the movement in an arbitrary direction is entirely ruled out. As for residing under the inner horizon, a falling body there is able to move along a world line away from the singularity. Using the Penrose diagram, it is possible to draw the pass of an astronaut who falls into a rotating black hole.

In the region confined by the inner horizon, the astronaut can change the direction of the own movement by, say, switching on the engines of his spaceship. At a velocity not exceeding that of light (that is, moving along the trajectory that forms an angle of less than 45° with the vertical), the astronaut can deviate from the singularity and even depart from it to eventually find himself in another spacetime.

Although the astronaut cannot return to our world from the black hole whereto he has fallen, it does not mean that he is unable to get out from the hole "somewhere" else, perhaps, in some other Universe. The full distribution of

Penrose makes it possible to represent an infinite number of universes of past and future. Getting into and out of rotating black holes, the astronaut can endlessly travel from one Universe to some other but only towards the "Universe of the future". Once entering a next black hole, he cannot come back to his own Universe and meet his coevals there.

Since each Universe implements its own way to translate Reality, we will be able to change between the forms of Reality representations; moreover, we mean here the switching between different *forms of consciousness* (in a Buddhist interpretation) because the Reality is unique. The fact that physicists assume the origin of our Universe as the Big Bang is in its essence the beginning of the translation of a certain Reality content into a particular spacetime form.

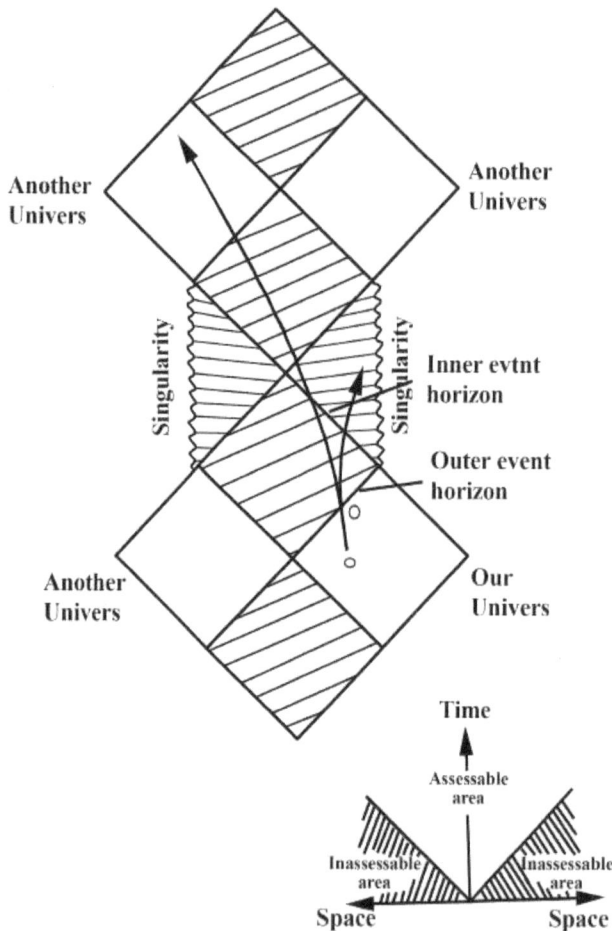

Fig. (8.2). Penrose diagram for a rotating black hole.

Analysis of the optical effects in the neighborhood of a black hole allowed German astrophysicists to design a specialized code for visualizing the movement of an observer around this mysterious object. Their paper was published in the Physics Reports [169]; the program is available on the site of Thomas Müller (one of the participants of this study).[16]

Fig. (**8.3**) (a–p) shows the main phases of passing through the outer and inner horizons of a black hole.

At the inner horizon (Fig. **8.3 g**), a bright radiant dot appears as a result of gravitational focusing, which is the image of the outer Universe reflected by the gravity repulsion of singularity and comprising the infinitely accelerated history of our entire Universe. With approaching the inner horizon, the extension of the black hole is changed for its collapse.

It looks as if you go away from the black hole but it is not so; the free fall into the black hole continues. As for the collapse, it results from relativistic radiation, focusing your eye forward. This radiation determines an increase in the irradiance of the outer Universe with a blue shift around the black hole.

At the inner horizon of the black hole, we encounter one more infinitely bright flash of light (outside of the "white hole"); this flash comprises the entire history of the "future Universe"[17] (Fig. **8.3 m**).

Now you enter a new Universe giving a farewell look towards the white hole you leave (Fig. **8.3 p**).

Designations:

Color	Area
Green	Stable circular orbits
Yellow	Unstable circular orbits
Orange	Absent circular orbits
Red lines	Horizons
Red	Between horizons

The clocks on the schemes show the time of a contact of internal horizon in seconds and define the sequence of shots (the falling from the outer to inner horizon takes 20 seconds in its own frame of reference for a black hole with a mass of 5 million solar masses).

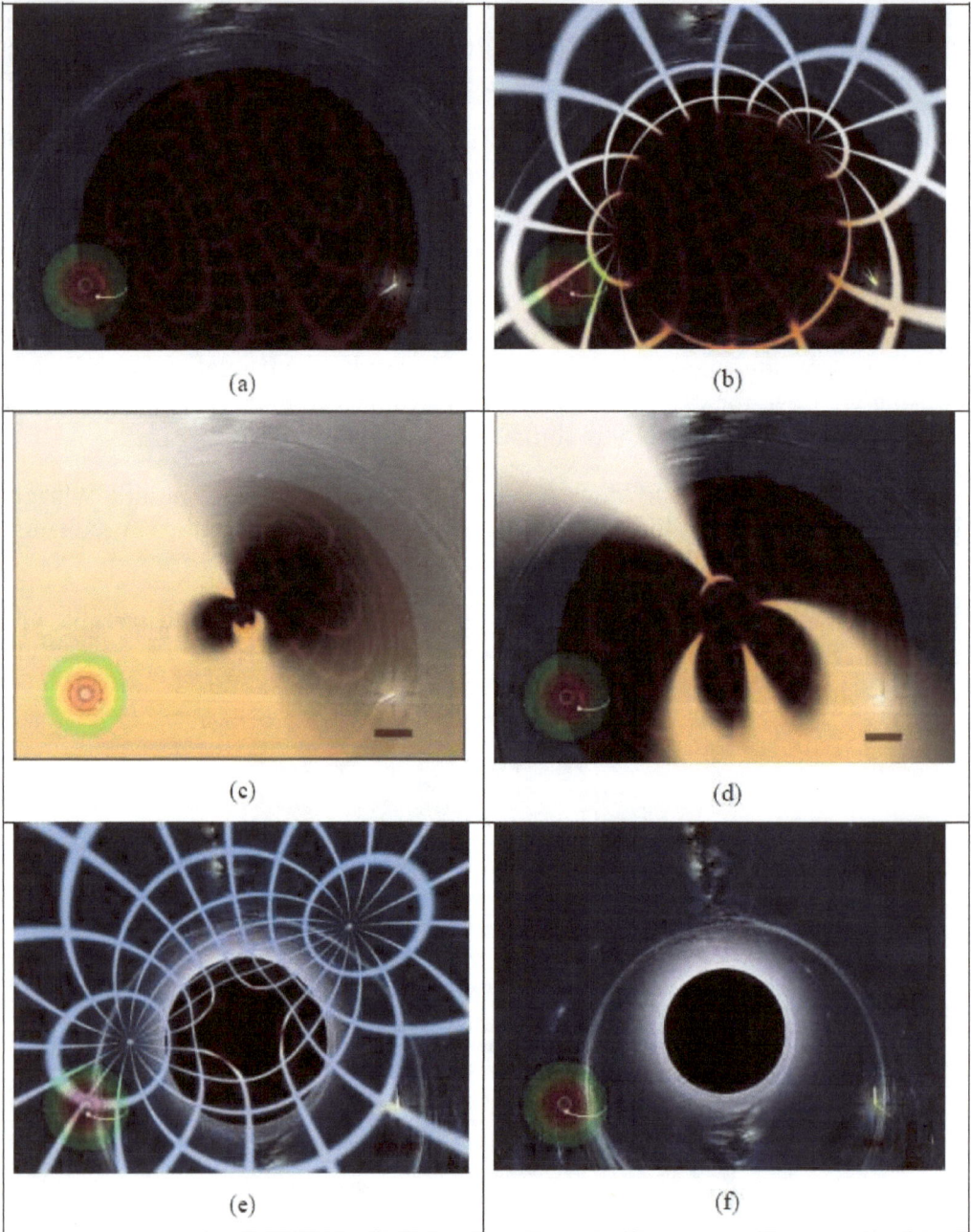

(a) (b) (c) (d) (e) (f)

(Fig. 8.3) contd.....

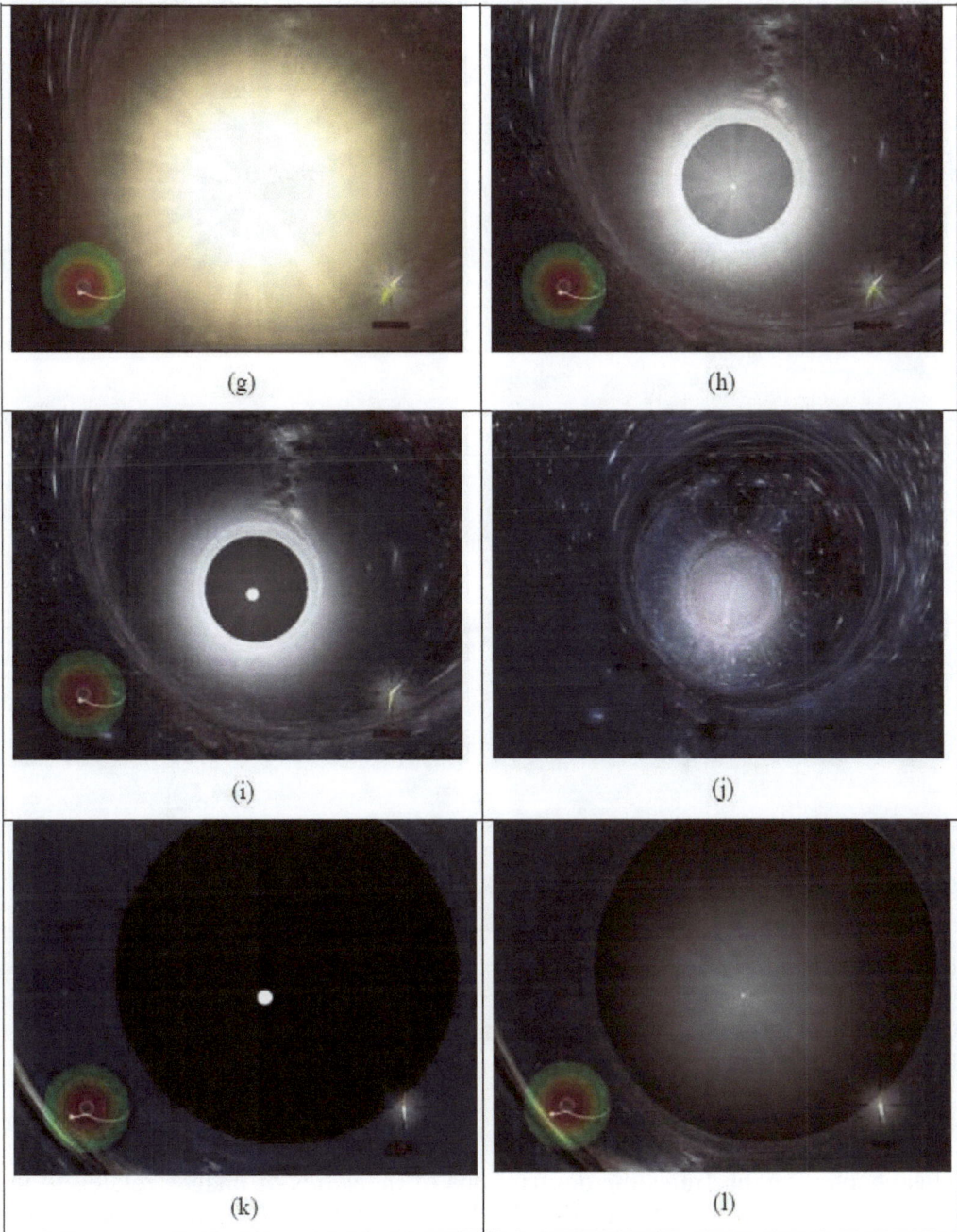

(g)

(h)

(i)

(j)

(k)

(l)

(Fig. 8.3) contd.....

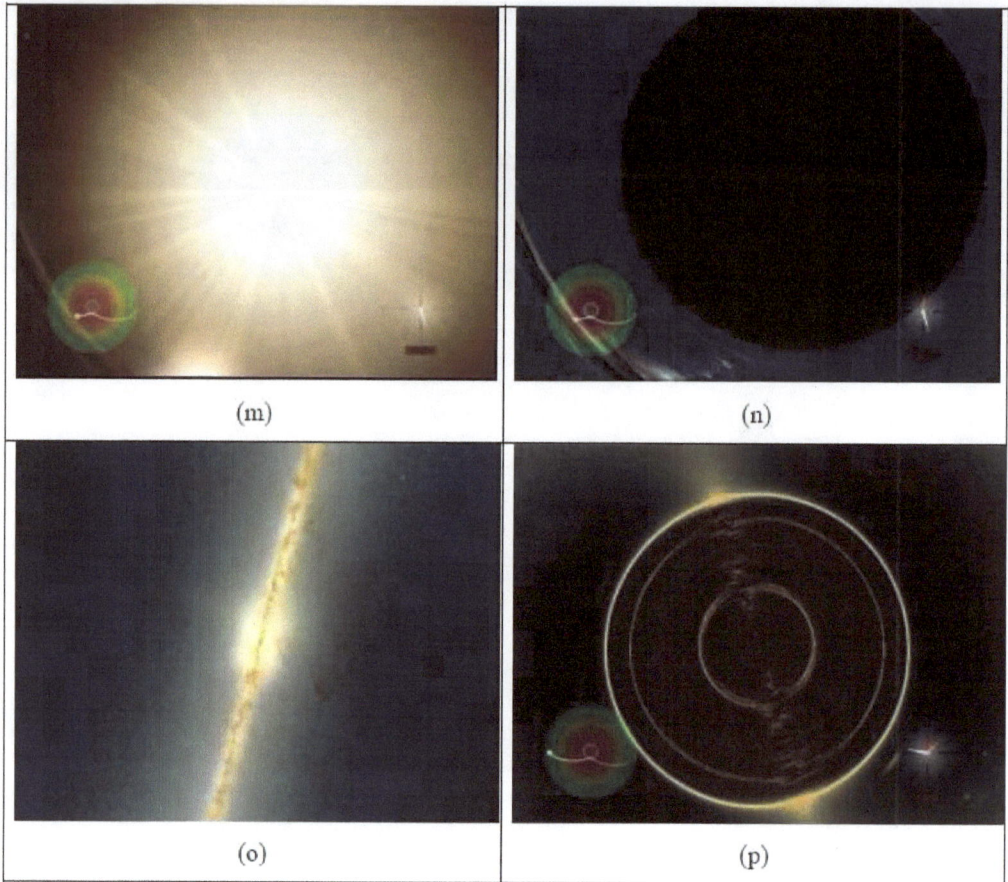

Fig. (8.3). Going through a wormhole.

Thus, not only metaphorical coincidence, but also literal ones are evident from this comparison. The movie *Black Holes: The Other Side of Infinity* (produced by the Denver Museum of Nature and Science and NASA Education and Public Outreach group) gives a fuller impression of this computer visualization of falling into a black hole.

CONCLUSION

This chapter demonstrates the descriptions of the series of images related to the process of dying and entering the Bardo state from the Buddhist book Bardo Thodol and the physical processes hypothetically taking place during the passage through a black hole in the cosmological model of wormhole suggest many amazing analogies. The parallels between the postulates of Buddhist philosophy, asserting that "everything is consciousness", and the Bohm–Pribram–Talbot holographic model of the Universe, regarding everything as the projection of an implicit "hologram", are not accidental. Two ways of modeling reality in physics are described, for example, with the help of the language of wave functions in Hilbert spaces and the language of classical physics as spacetime processes. As has emerged, these representations are comparable to the mental processes of the preconscious (unconscious) level, while the worldview generated by a discrete sign-based conscious process corresponds to the second method of description.

NOTES

[1] Nagarjuna was a great Indian religious philosopher and thinker; he lived in the 3rd–2nd centuries BC.

[2] For details, see *Matematika. Bol'shoi entsiklopedicheskii slovar'* (Mathematics. Large Encyclopedic Dictionary) [136], p. 44.

[3] In particular, the experiments by Leontiev [140] on development of skin light sensitivity have convincingly demonstrated that an abiotic stimulus (*i.e.*, biologically neutral stimulus)—a light beam—can be sensed by the human skin only if it has a signaling function that informs about a biologically significant event (in this case, electric shock).

[4] Quantum theory has a convenient method for description and computation of the processes underlying the interaction of particles; this method is based on the Feynman diagrams. In these diagrams, a graphical scheme corresponds to each physical process and a line corresponds to each particle involved in the process. The lines in a Feynman diagram can describe the propagation of both particles and antiparticles: the direction of arrows on the lines for antiparticles is opposite to that for the particles.

[5] In addition to the harmonic Fourier transform, other methods of signal representation are also used, such as Walsh, Bessel, and Haar functions and Chebyshev, Laguerre, and Legendre polynomials.

[6] Nirvana within the limits of Samsara is Sunyata, *i.e.*, emptiness or nothingness.

[7] In fact, we can determine the accuracy of measurement in an experiment by choosing either the width of time window, Δt (through which time is "translated" to our consciousness), or the width of spectrum window $\Delta \omega$ since they are

connected through the Fourier transform and an increase in one variable leads to a decrease in the other.

[8] Dzogchen (rdzogschen in Tibetan), or Great Perfection, is the highest and most profound teaching of the Shakyamuni Buddha.

[9] Bardo is a Tibetan word meaning the "transition" to or interval between the completion of a situation and beginning of the next situation.

[10] Rigpa (vidya in Sanskrit) is the Dzogchen term meaning a mind free of ignorance and dualistic perception and forms the background of the thinking ability.

[11] The Sanskrit word dharmata (choniy in Tibetan) means the intrinsic nature of everything, the essence of things as they are.

[12] The spacetime region in which the curvature of spatiotemporal continuum becomes infinite or discontinues or the metric has other pathological features that are physically uninterpretable.

[13] Its geometry was independently described by Reissner (1916) and Nordström (1918).

[14] The general theory of relativity does not deny the existence of such tunnels. It is necessary for the existence of a passable wormhole that is filled by an exotic matter creating strong gravitational repulsion that prevents its collapse.

[15] A technique of "bringing consciousness into the space of unborn Rigpa".

[16] The paper with the theoretical substantiation is available in arXiv.org and the video, at http://casa.colorado.edu/~ajsh/insidebh/schw.html.

[17] This happens when the pass changes the direction of time but this is the *past* for the new Universe.

AFTERWORD

"Is there anything of which one can say, 'Look! This is something new?' It was here already, long ago; it was here before our time."
Ecclesiastes 1:10

Мыслам сотни лет, а свежи.
Чуда нет - грехи все те же. c_1

The thoughts are centuries old, still fresh they are. The sins remain the same and it's no miracle there.
Midnight thoughts

We started our narration by mentioning the notorious crisis in physics [169–172] as a case suggesting a revision of our scientific foundations. Crises every now and then take place in any area of human activity when the existing behavioral algorithms and the line of action appear inadequate to the current situation because of changed conditions or give unsatisfactory results. Indeed, the geocentric model by Ptolemy gave place to the heliocentric model, which described well the movement of planets with concurrent simplicity and clearness. Strange it may seem but accuracy here is a secondary factor since it can be usually improved by making the model more complex. Followers of Ptolemy actually made attempts of this kind. This means that the selection of a model, *i.e.*, a useful metaphor, is the subject of agreement, as Poincare stated [18]. The situation is somewhat more difficult if we need to revise our views of reality in general rather than correct a particular model.

The human consciousness is not omnipotent and we have the right to make mistakes; however, it should not be overused. The historical artifacts, such as Egyptian pyramids, murals of Native Americans, megalithic constructions, and so on, have been long discussed. In our life, we have not encountered the need to create something of this kind and, thus, we tend to see modern vehicles, spaceships, or astronauts in spacesuits, that is, everything that resemble the known objects, in these ancient drawings. We reduce the observed phenomena to conventional forms and then make conclusions that most likely make no sense.

Turning to the problems in modern physics, we see that, strange as it may seem, the situation here is very similar to the cases described above. We ventured into the area of reality that we can judge based on instrumental data only, as if we record the "message of civilization" that is at a completely different level of development, while the unconscious "suggests" us the model of the world that was targeted to satisfy the simple needs of *Homo sapiens*. Presumably, the look on reality *via* an electron microscope demands some other approaches to adequately understand the situation; however, we cannot control the unconscious and the algorithms of its operation.

It is necessary to state that our worldview and its scientific interpretation are based on an informed analysis of the model presented to us by the unconscious. This is an object-based model because of several reasons. Presumably, the "technical" side of such processing of the first signal system data has emerged to be the most appropriate for solving the problems set by evolution. In addition, distinguishing something and keeping it in the focus of attention in this way of partitioning the integral picture into details could appear the most efficient approach to achieving the goal in human practice. Modern civilization is the product of the creative activity of individuals belonging to our species, especially, from the technological point of view. The advances in the manufacture of various objects according to the acquired algorithms are impressive and the changes in the lifestyle, especially, during the last half-century, are cardinal.

The object-based model also has made it possible to simply and illustratively construct the cause-and-effect relations *via* the concept of direct interaction between objects. In addition, once the maximally possible velocity of signal propagation was postulated, the overall world logically fell apart into the interdependent portion and everything else. This could be regarded as a direct proof that independent events do exist and this inference was the scientific basis for the correctness of reality representation as a set of self-sufficient objects.

Once we had partitioned everything, it emerged that it was necessary to introduce the relations between objects; otherwise, our model literally "fell into pieces". It was logical to add the concept of the field of any given nature and then to try to consider it in the format of objects, namely, as the particles that transferred a given type of interaction. However, the description of objects requires the notions of space and time as their receptacle; moreover, time was also included in geometry as a full-fledged dimension. In addition, the desire to create a comprehensive model of the world has led to the attempts to unite all existing metaphors in a single theory of everything. This failed to fit a four-dimensional space, giving birth to multidimensional space with folded dimensions. There is no end in sight for this chain of objectivation although common sense prompts that any metaphor must have its limits. Thus, consciousness appeared to be a puppet manipulated by our unconscious.

We have come to believe that reality is actually representable as a set of objects but this is not the main point. The main problem is that our faith is blind in all respects. Any attempts to analyze the existing paradigm are absent in science; moreover, the meaning of the term "object" is not discussed in physics at all. Textbooks tell that an electron can behave as a particle or a wave, which is followed by mathematical equations that correctly describe the results of experiments. Actually, the problem of our modeling of reality is swept under the carpet of mathematics. The trailblazers in quantum mechanics desired to understand the content of the new world they discovered; however, everything boiled down to learning the texts approved by experts a century later. The current system of education is not aimed at teaching to think but rather on the contrary with the goal to incorporate a student into the existing social institutions dictating a certain set of traditional prejudices.

In this book, we have attempted to substantiate the statement that man lives in a virtual world constructed by his unconscious, which is confirmed in the current studies in the area of psychophysiology. Presumably, this will raise an objection of the senior generation because of the formed belief in the absence of alternatives to "objective materialism". However, the generation that grew up in the recent 20 years and was "vaccinated" by computer technologies will most likely take it calmer.

It is difficult to object anything to the statement that we can perceive reality only *via* our sensory organs and that any other world is unavailable to us. In particular, Schrödinger in his *Mind and Matter* writes, "…The world is given to me only once, not one existing and one perceived. Subject and object are only one" [146]. This means that it is not logical to exclude oneself from the world assuming that this will make the world "objective" because this is groundless. However, this also does not mean that our perceptions (sensations) are groundless too. "Admittedly our sense perceptions constitute our sole knowledge about things. This objective world remains a hypothesis, however natural. If we do adopt it, is it not by far the most natural thing to ascribe to that external world, and not to ourselves, all the characteristics that our sense perceptions find in it?" [146].

The world exists and, first and foremost, it is characterized by its integrity. The experiments in quantum physics have convincingly demonstrated the wholeness of reality; correspondingly, our partitioning of the world into objects and the introduction of boundaries are conventional. First and foremost, this is true for the concepts, such as matter and consciousness, the opposition of which has actually arrested the development of our insight into the world and ourselves. If the principle of locality in physics is false, it just does not exist. According to Schrödinger, "It is very difficult for us to take stock of the fact that the localization of the personality, of the conscious mind, inside the body is only symbolic, just an aid for practical use" [146].

As is shown above, our social existence makes us to agree with our individual mental maps. This unification at a conscious level is necessary to present the personal ego, *i.e.*, to express our individuality in the way that we are understood. However, the reality is what truly unites us since we are intrinsically incorporated into this reality most likely at the level of the unconscious. This approach could promote the solution of the problems described in the holographic models of the Universe and consciousness [115, 116].

One might wonder what psychology has to do with physics since the latter studies reality. However, an observer, who records the results of an experiment and acts there as an unerring tool, appears as early as the relativity theory. As for quantum mechanics, it became apparent that the reduction of the wave function takes place only in the event of observation. Measurement is regarded as an irreversible process that reduces the state of a quantum system to reality. However, if we are consistent in this case, the experimenter should be also regarded as a system included in the study. This is an argument favoring the fact that physicists need to represent the operation principles of this "device" and the range of reliability of the results thus obtained. Eventually, the interpretation of quantum mechanics becomes epistemic rather than ontological, *i.e.*, this problem is associated with the ways in which the information about it is acquired rather than with the nature of reality [173].

Physicists have attempted to resolve the problem of wave function reduction by proposing a model of "branched" Universe (the Everett–Wheeler worlds); in this model, each existing probability of a superposition of states is implemented in one of the parallel Universes. Leaving aside the physical meaning of this theory, which looks unexplainable, let us analyze the content of this theory in terms of logic. The probability amplitude is the conceivable results of observation of a quantum system before its translation (wave function reduction during the event of measurement) into an object-based model of our consciousness. This is interpretable as the existence of the superposition of possible outcomes

(Universes) before the event of measurement with the subsequent passage to a particular "branch" in reality after this measurement. As we have mentioned earlier, this blend of the classical and quantum models of reality has no sense at all. The Total Reality is the Absolute comprising all variants of what is translatable into the system of consciousness and its elements (subsystems) are describable only with the atemporal category of possibilities. The classical probability in terms of the object-based model is the mathematical limit for the outcome of the process of testing with the number of tests going to infinity (limi→ ∞pi), and, correspondingly, requiring an infinite time. In this case, there is no place for interference because the results do not exist simultaneously and the tests are independent. The probability amplitude corresponds to the representation of the system in Hilbert's space, in which the concept of time is absent (the harmonic is infinite). Correspondingly, it is inappropriate to mix two forms of reality representation—sequential spacetime and quantum integral ones—within the same model. A number of scientists believe that the hypothesis of the Multiverse is more philosophic rather than scientific since it is unfalsifiable, that is, it cannot be refuted with the help of a scientific extract, which is an indispensable part of the scientific method.

We continue to spend tremendous efforts and money to get new data but do not envisage the direction of our way. The situation has arisen when some theories, such as the string theory, appear and are further developed; however, this theory even cannot be experimentally verified [174]. Another example is that according to the Big Bang theory, everything has emerged from "nothing" but this requires according to the physical "bookkeeping" that the total of all qualities must be zero. This refers to energy, momentum, symmetry, mass, and so on but we have already discovered that this is not so, while it is completely vague how the recently discovered "dark" components could be compensated. In addition, it follows from the Big Bang theory at least at a quantum level that everything in the world must be entangled and this fact most likely cannot be omitted from consideration.

As for the notions of space and time, we may say that the compatibility of quantum mechanics and spacetime concepts has not been achieved so far. Moreover, the experiments with EPR pairs have showcased the absence of any independent events in a conventional sense. This fact demolished not only the logic of causal relations but also the postulate on the limit velocity at least for the transmission of the state of a quantum system. This could become the trigger for a serious revision of the science foundation but, as we see it, it did not.

The ideas described here fall beyond the scope of physics per se and we hope that they will give a boost to the research into the neighboring areas aiming to verify the made suggestions. The considered paradigm may well be used as an interdisciplinary in different fields of science forming the background that unites these areas just because human perception underlies any science. This gives the possibility to look from a united position but from another perspective, to look on

the problems existing in different areas of knowledge, which may be helpful in their resolution.

Indeed, we do not call for rejecting the "spacetime" terminology in the modeling of reality. We just try to attract attention to the fact that this approach is not universal and has its range of applicability. Note that the meaning of the notions is comprehensible only as *the meaning of a subject.* The attempt to understand the meaning of reality with the alienation of the subject, "discarding" the subject from the scientific paradigm, is analogous to the alienation of the Cheshire cat smile from the animal. Modern science inexplicitly ascribes a real existence to the notions themselves as a "material" manifestation of the "virtual" world "existing" without the subject. Thus, the very notions substitute the direct Reality, which can be only of a subject-based nature.

References

[1] T. Metzinger, *The Ego Tunnel. The Science of the Mind and the Myth of the Self.,* Basic Book: New York.

[2] J. Neumann, *Mathematical Foundations of Quantum Mechanics.,* Princeton University Press.

[3] X. Ma, J. Kofler, and A. Zeilinger, "Delayed-choice gedanken experiments and their realizations", *Rev. Mod. Phys.,* vol. 88, no. 1, p. 015005.[http://dx.doi.org/10.1103/RevModPhys.88.015005]

[4] A. Petersen, The philosophy of Niels Bohr.*Niels Bohr: A Centenary Volume.,* Harvard University Press: Cambridge, Mass., .

[5] A.A. Tyapkin, Principle of relativity.*Collection of Works on the Special Theory of Relativity.,* Atomizdat: Moscow, . [in Russian]

[6] S.P. Suprun, and A.P. Suprun, *Computers: Classical, Quantum and Others.,* Bentham Science Publishers: Dubai, .

[7] R. Penrose, *Shadows of the Mind: A Search for the Missing Science of Consciousness.,* Oxford University Press: New York, .

[8] B. Russell, *The Problems of Philosophy.,* Oxford University Press, .

[9] S. Vivekananda, *Jnana Yoga.,* Ramakrishna-Vivekananda Center: New York, .

[10] "The Collected Papers of Albert Einstein", *Princeton University Press,* .

[11] N. Wiener, *I Am a Mathematician, the Later Life of a Prodigy.,* The MIT Press, .

[12] D. Bohm, *The Special Theory of Relativity.,* W.A. Benjamin, Inc., .

[13] L. Von Bertalanffy, General system theory: A critical review.*General System Theory: Foundations, Development, Applications.,* George Braziller: New York, .

[14] S. Suprun, and A. Suprun, Models of reality in physics[http://dx.doi.org/10.13140/RG.2.1.2728.5922]

[15] S. Suprun, "The experiments with EPR pairs: conclusions", [http://dx.doi.org/10.13140/RG.2.1.2728.5922]

[16] D. Deutsch, and A. Ekert, "Beyond the quantum horizon", *Sci. Am.,* vol. 307, no. 3, pp. 84-89.[http://dx.doi.org/10.1038/scientificamerican0912-84] [PMID: 22928266]

[17] H. Poincare, *The Value of Science.,* Dover Publication, Inc.: New York, .

[18] H. Poincare, *Science and Hypothesis.,* Walter Scott Publishing Co., .

[19] H. Poincare, *The Foundations of Science: Science and Hypothesis, The Value of Science, Science and Method.,* The Science Press, New York Garrison: New York, . [EBook #39713]

[20] L.K. Palmer, and J.A. Palmer, *The Ultimate Origins of Human Behavior.,* Allyn and Bacon: Boston, .

[21] W. Heisenberg, *Physics and Philosophy.,* Harper and Brothers, .

[22] *The Theory of Sets: From Cantor to Cohen.,* vol. 16, URSS: Moscow, .V. Boss, Lectures in Mathematics [in Russian]

[23] H. Poincare, *Mathematics and Science: Last Essays..* Dover Publications, 1963.

[24] S. Augustine, *Confessions.,* Hackett Publishing, .

[25] A. Einstein, B. Podolsky, and N. Rosen, "Can quantum-mechanical description of physical reality be considered complete?", *Phys. Rev.,* vol. 47, no. 10, pp. 777-780.[http://dx.doi.org/10.1103/PhysRev.47.777]

[26]   A.R. Luria, *The Foundations of Neuropsychology.*, Akademiya: Moscow, . [in Russian]

[27]   A. Aspect, "Proposed experiment to test the nonseparability of quantum mechanics", *Phys. Rev. D Part. Fields,* vol. 14, no. 8, pp. 1944-1951.[http://dx.doi.org/10.1103/PhysRevD.14.1944]

[28]   A. Aspect, J. Dalibard, and G. Roger, "Experimental test of Bell's inequalities using time-varying analyzers", *Phys. Rev. Lett.,* vol. 49, no. 25, pp. 1804-1807.[http://dx.doi.org/10.1103/PhysRevLett.49.1804]

[29]   D. Bouwmeester, A. Ekert, A. Zeilinger, Ed., *The Physics of Quantum Information.*, Springer, .[http://dx.doi.org/10.1007/978-3-662-04209-0]

[30]   G. Greenstein, A.G. Zajonc, Ed., *The Quantum Challenge.*, Jones and Bartlett Publishers, Inc., .

[31]   J. Preparata, *An Introduction to a Realistic Quantum Physics.*, World Scientific: Singapore, .[http://dx.doi.org/10.1142/5111]

[32]   P. Grangier, G. Roger, and A. Aspect, "Experimental evidence for a photon anti-correlation effect on a beam splitter", *Europhys. Lett.,* vol. 1, p. 173.[http://dx.doi.org/10.1209/0295-5075/1/4/004]

[33]   M. Kumar, *Quantum: Einstein, Bohr, and the Great Debate about the Nature of Reality.*, Icon Books, .

[34]   M.A. Nielsen, and I.L. Chuang, *Quantum Computation and Quantum Information.*, Cambridge University Press, .

[35]   B.I. Spasskii, and A.V. Moskovskii, "Nonlocality in quantum physics", *Усп. Физ. Наук,* vol. 142, no. 4, p. 599.[http://dx.doi.org/10.3367/UFNr.0142.198404c.0599]

[36]   G.N. Afanas'ev, "The old and new problems in the theory of Aharonov–Bohm effect, Fiz. Element. Chast. Atom", *Yadra,* vol. 21, no. 1, p. 172.

[37]   R. Penrose, *The Emperor's New Mind.*, Oxford University Press, .[http://dx.doi.org/10.1093/oso/9780198519737.001.0001]

[38]   E.P. Wigner, Remarks on the mind–body question.*The Scientist Speculates.*, Heinemann: London, .

[39]   G. Moruzzi, and H.W. Magoun, "Brain stem reticular formation and activation of the EEG", *Electroencephalogr. Clin. Neurophysiol.,* vol. 1, no. 1-4, pp. 455-473.[http://dx.doi.org/10.1016/0013-4694(49)90219-9] [PMID: 18421835]

[40]   C.F. Jacobsen, *Functions of Frontal Associations in Monkeys.*, Monographs on Comparative Psychology, .

[41]   P.K. Anokhin, " Characteristics of the afferent apparatus of a conditioned reflex and its importance for psychology", *Vopr. Psikhol,* vol. 6, no. 16, .

[42]   P.H. Lindsay, and A.D. Norman, *Human Information Processing (An Introduction to Psychology).*, Academic Press: New York, London, .

[43]   A. Stevens, *Archetypes: A Natural History of the Self.*, William Morrow: New York, .

[44]   P. van Lommel, R. van Wees, V. Meyers, and I. Elfferich, "Nahtoderfahrung bei Überlebenden eines Herzstillstands: eine prospektive Studie in den Niederlanden", *Lancet,* vol. 358, p. 2045.

[45]   H.E. Puthov and R. Targ, A perceptual channel for information transfer over kilometer distances: Historical perspective and recent research, *IEEE,* 64 (3), 329, 1976.

[46]   V.F. Petrenko, "Psychosemantic aspects of the worldview of a subject", *Psikhol. Zh. Vyssh. Shkoly Ekon.,* vol. 2, no. 2, p. 3.

[47]   R.O. Jakobson, A few remarks on Pierce, pathfinder in the science of language, In: *The Framework of Language (Michigan Studies in the Humanities)*, Horace H. Rackham School of Graduate Studies, 31, 1980.

[48]   N.A. Bernshtein, *Physiology of Movements and Activity.*, Nauka: Moscow, . [in Russian]

[49]   A.D. Logvinenko, *Measurements in Psychology: Mathematical Foundations,* .

[50]   A.R. Luria, *Language and Consciousness.,* Rostov-on-Don, . [in Russian]

[51]   A.P. Suprun, N.G. Yanova, and K.N. Nosov, *Metapsychology.,* Relativistic Psychology. Quantum Psychology. Psychology of Creativity: Moscow, . [in Russian]

[52]   V.F. Petrenko, *Psychosemantics Bases.,* Moscow, . [in Russian]

[53]   D.I. Blokhintsev, *Principle Points of Quantum Mechanics.,* Nauka: Moscow, . [in Russian]

[54]   E. Schrödinger, *What is Life?,* Cambridge Press, .

[55]   V.A. Bufeev, *Who and How Created the Theory of Relativity: The History of Creation of Development of Understanding,* .

[56]   R. Penrose, *The Road to Reality.,* Jonathan Cape: London, .

[57]   V.A. Vasil'ev, *Introduction to Topology.,* Fazis: Moscow, . [in Russian]

[58]   S. Kobayashi, and K. Nomizu, *Foundations of Differential Geometry.,* vol. Vol. 1, Interscience Publishers: New York, .

[59]   V.A. Rokhlin, and D.B. Fuks, *The Introductory Course to Topology.,* Nauka: Moscow, . [in Russian]

[60]   A.N. Whitehead, *The Concept of Nature.,* Cambridge, .

[61]   B. Zeigarnik, "Das Behalten erledigter und unerledigter Handlungen", *Psychol. Forsch.,* vol. 9, p. 1.

[62]   S.S. Korsakov, *Selected Works.,* Gos. Izd. Med. Lit: Moscow, . [in Russian]

[63]   S. Hawking, *A Brief History of Time.,* Bantam Books, .[http://dx.doi.org/10.1063/1.2811637]

[64]   C.G. Jung, *Psychology of the Unconscious.,* Moffat, Yard and Company: New York, .

[65]   N. Gisin, *Quantum Chance, Nonlocality, Teleportation and Other Quantum Marvels.,* Springer, .

[66]   M. Giustina, M.A.M. Versteegh, S. Wengerowsky, J. Handsteiner, A. Hochrainer, K. Phelan, F. Steinlechner, J. Kofler, J.Å. Larsson, C. Abellán, W. Amaya, V. Pruneri, M.W. Mitchell, J. Beyer, T. Gerrits, A.E. Lita, L.K. Shalm, S.W. Nam, T. Scheidl, R. Ursin, B. Wittmann, and A. Zeilinger, "Significant-Loophole-Free Test of Bell's Theorem with Entangled Photons", *Phys. Rev. Lett.,* vol. 115, no. 25, p. 250401.[http://dx.doi.org/10.1103/PhysRevLett.115.250401] [PMID: 26722905]

[67]   L.K. Shalm, E. Meyer-Scott, B.G. Christensen, P. Bierhorst, M.A. Wayne, M.J. Stevens, T. Gerrits, S. Glancy, D.R. Hamel, M.S. Allman, K.J. Coakley, S.D. Dyer, C. Hodge, A.E. Lita, V.B. Verma, C. Lambrocco, E. Tortorici, A.L. Migdall, Y. Zhang, D.R. Kumor, W.H. Farr, F. Marsili, M.D. Shaw, J.A. Stern, C. Abellán, W. Amaya, V. Pruneri, T. Jennewein, M.W. Mitchell, P.G. Kwiat, J.C. Bienfang, R.P. Mirin, E. Knill, and S.W. Nam, "Strong Loophole-Free Test of Local Realism", *Phys. Rev. Lett.,* vol. 115, no. 25, p. 250402.[http://dx.doi.org/10.1103/PhysRevLett.115.250402] [PMID: 26722906]

[68]   J.A. Wheeler, W.H. Zurek, Ed., *Quantum Theory and Measurement.,* Princeton University Press, .

[69]   J.S. Tang, Y.L. Li, X.Y. Xu, G.Y. Xiang, C.F. Li, and G.C. Guo, "Realization of quantum Wheeler's delayed-choice experiment", *Nat. Photonics,* vol. 6, no. 9, pp. 600-604.[http://dx.doi.org/10.1038/nphoton.2012.179]

[70]   A. Peruzzo, P. Shadbolt, N. Brunner, S. Popescu, and J.L. O'Brien, "A quantum delayed-choice experiment", *Science,* vol. 338, no. 6107, pp. 634-637.[http://dx.doi.org/10.1126/science.1226719] [PMID: 23118183]

[71]   S.S. Roy, A. Shukla, and T.S. Mahesh, "NMR implementation of a quantum delayed-choice experiment", *Phys. Rev. A,* vol. 85, no. 2, p. 022109.[http://dx.doi.org/10.1103/PhysRevA.85.022109]

[72]   E. Schrödinger, "Die gegenwärtige Situation in der Quantenmechanik", *Nsturwissenschatten,* vol. 23, p. 844.[http://dx.doi.org/10.1007/BF01491987]

[73] A. Peres, "Delayed choice for entanglement swapping", *J. Mod. Opt.,* vol. 47, no. 2-3, pp. 139-143.[http://dx.doi.org/10.1080/09500340008244032]

[74] X. Ma, S. Zotter, J. Kofler, R. Ursin, T. Jennewein, Č. Brukner, and A. Zeilinger, "Experimental delayed-choice entanglement swapping", *Nat. Phys.,* vol. 8, no. 6, pp. 479-484.[http://dx.doi.org/10.1038/nphys2294]

[75] V. Jacques, E. Wu, F. Grosshans, F. Treussart, P. Grangier, A. Aspect, and J.F. Roch, "Experimental realization of Wheeler's delayed-choice gedanken experiment", *Science,* vol. 315, no. 5814, pp. 966-968.[http://dx.doi.org/10.1126/science.1136303] [PMID: 17303748]

[76] X.S. Ma, J. Kofler, A. Qarry, N. Tetik, T. Scheidl, R. Ursin, S. Ramelow, T. Herbst, L. Ratschbacher, A. Fedrizzi, T. Jennewein, and A. Zeilinger, "Quantum erasure with causally disconnected choice", *Proc. Natl. Acad. Sci. USA,* vol. 110, no. 4, pp. 1221-1226.[http://dx.doi.org/10.1073/pnas.1213201110] [PMID: 23288900]

[77] F. Kaiser, T. Coudreau, P. Milman, D.B. Ostrowsky, and S. Tanzilli, "Entanglement-enabled delayed-choice experiment", *Science,* vol. 338, no. 6107, pp. 637-640.[http://dx.doi.org/10.1126/science.1226755] [PMID: 23118184]

[78] C.O. Alley, O. Jakubowicz, and W.C. Wickes, *2nd International Symposium on Foundations of Quantum Mechanics in the Light of New Technology,* Physical Society of Japan: Tokyo, .

[79] R. Penrose, *Fashion, Faith, and Fantasy in the New Physics of the Universe.,* Princeton University Press, .

[80] D.P. Lindorff, *Pauli and Jung: The Meeting of Two Great Minds.,* Quest Books, .

[81] W. Arntz, B. Chasse, and M. Vicente, *What the Bleep Do We Know? Discovering the Endless Possibilities for Altering Your Everyday Reality.,* Health Communications, Inc.: Deerfield Beach, .

[82] V.I. Lebedev, *Genii in the Mirror of Psychology.,* Moscow, . [in Russian]

[83] P.A. Florensky, *Mnimosti in Geometry.,* Nauka: Moscow, . [in Russian]

[84] N.A. Vasil'ev, *On Particular Judgments, Triangle of Opposites, and the Law of Excluded Fourth.,* Tipo-litografiya Universiteta: Kazan, . [in Russian]

[85] R. Feynman, R.B. Leighton, and M. Sands, *The Feynman Lectures on Physics*.http://www.feynmanlectures.caltech.edu/III_toc.html[http://dx.doi.org/10.1063/1.3051743]

[86] *Equations of Mathematical Physics.,* vol. 11, Editorial URSS: Moscow, .V. Boss, Lectures in Mathematics [in Russian]

[87] Aristotle, *Metaphysics.,* Green Lion Press, .

[88] R.W. Clark, *Albert Einstein: The Life and Time.,* World Publishing Company, .

[89] P.K. Anokhin, *Outline of the Physiology of Functional Systems.,* Moscow, . [in Russian]

[90] A.R. Luria, *Higher Cortex Functions and Their Disturbance in Local Brain Injuries.,* Moscow, . [in Russian]

[91] E.A. Singer, *Mind as Behavior,* Colubus: R.G. Adams, .

[92] E.A. Singer, *Experience and Reflection.,* University of Pennsylvania Press: Philadelphia, .[http://dx.doi.org/10.9783/9781512806960]

[93] A. Rosenblueth, N. Wiener, and J. Bigelow, "Behavior, purpose and teleology", *Philos. Sci.,* vol. 10, no. 1, pp. 18-24.[http://dx.doi.org/10.1086/286788]

[94] A. Rosenblueth, and N. Wiener, "Purposeful and non-purposeful behavior", *Philos. Sci.,* vol. 17, no. 4, pp. 318-326.[http://dx.doi.org/10.1086/287107]

[95] G. Sommerhoff, *Analytical Biology.,* Oxford University Press: London, .

[96] C.W. Churchman, *The System Approach.*, Delacorte Press: New York, .

[97] R.L. Ackoff, and F.E. Emery, *On Purposeful Systems: An Interdisciplinary Analysis of Individual and Social Behavior as a System of Purposeful Event.*, Aldine Transactions, .

[98] Yu.I. Aleksandrov, Introduction to systems psychophysiology [in Russian]

[99] B. Russell, *A History of Western Philosophy and its Connection with Political and Social Circumstances from the Earliest Times to the Present Day.*, Simon and Schuster: New York, .

[100] C.E. Shannon, "A mathematical theory of communication", *Bell Syst. Tech. J.*, vol. 27, no. 3, pp. 379-423.[http://dx.doi.org/10.1002/j.1538-7305.1948.tb01338.x]

[101] T.A. Cowan, "Decision theory in law, science, technology", *Science,* vol. 140, no. 3571, pp. 1065-1075.[http://dx.doi.org/10.1126/science.140.3571.1065] [PMID: 17794895]

[102] G.A. Miller, E. Galanter, and K. Pribram, *Plans and the Structure of Behavior.*, Holt, Rinehart and Winston: New York, .[http://dx.doi.org/10.1037/10039-000]

[103] V.S. Stepin, *Theoretical Knowledge.*, Progress-Tradistiya: Moscow, . [in Russian]

[104] G. Sommerhoff, The abstract characteristics of living systems.*Systems Thinking.*, Penguin, .

[105] *Philosophical Encyclopedic Dictionary,* Sovetskaya entsiklopediya: Moscow, .

[106] R.J.O. Smith, and J. Stephenson, *Computer Simulation of Continuous Systems.*, Cambridge University Press, .

[107] M.D. Mesarović, *Foundations for a General Systems Theory,* .

[108] I.M. Kobozeva, *Linguistic Semantics.*, Editorial URSS: Moscow, . [in Russian]

[109] H. Harman, *Modern Factor Analysis.*, University of Chicago Press, .

[110] F.R. Gantmacher, *Theory of Matrices.*, Chelsea Pub. Co., .

[111] A.M. Dubrov, *Processing of Statistical Data by Principal Component Analysis.*, Moscow, . [in Russian]

[112] K.E. Tsiolkovsky, *Cosmic Philosophy.*, Editorial URSS: Moscow, . [in Russian]

[113] V.I. Vernadsky, *The Living Matter.*, Nauka: Moscow, . [in Russian]

[114] A.L. Chizhevsky, *The Cosmic Pulse of Life.*, Mysl: Moscow, . [in Russian]

[115] D. Bohm, *Wholeness and the Implicate Order.*, Routledge: London, .

[116] K. Pribram, *Languages of the Brain.*, Wadsworth Publishing: Monterey, Calif., .

[117] M. Talbot, *The Holographic Universe.*, Harper Perennial, .

[118] B. Lipton, *The Biology of Belief: Unleashing the Power of Consciousness.*, Matter and Miracles, .

[119] D.E. Durkheim, *Sociology and Its Scientific Domain,* .

[120] J. Piaget, *Selected Works on Psychology.*, Mezhdunar. Pedagog. Akad.: Moscow, . [in Russian]

[121] H. Wallon, *From Act to Thought.*, Moraes Editores: Lisbon, .

[122] V.F. Petrenko, and A.P. Suprun, "Consciousness and Reality in Western and Oriental Tradition. Relationship between human and universe", *Psychology in Russia: State of Art,* vol. 5, no. 1, p. 74.[http://dx.doi.org/10.11621/pir.2011.0006]

[123] S. Hodge, and M.J. Boord, *The Tibetan Book of the Dead.*, Godsfield Press Ltd., .

[124] A.H. Maslow, *Motivation and Personality.*, Harper & Row: New York, .

[125] S. Grof, *The Holotropic Mind: The Three Levels of Human Consciousness and How They Shape Our Lives.*, Harper One, .

[126] K. Wilber, *Integral Psychology.,* Shambhala: Boston.

[127] C. Tart, *Altered States of Consciousness.,* John Wiley & Sons Inc.: New York, London, Sydney, Toronto.

[128] R. Frager, and J. Fadiman, *Humanistic, Transpersonal, and Existential Psychology: C. Rogers, A. Maslow, and R. May.,* Praim-Evroznak: St. Petersburg. [in Russian]

[129] V.V. Kozlov, and V.V. Maikov, *Transpersonal Psychology. History, Origins, and Current State.,* Institute of Transpersonal Psychology: Moscow. [in Russian]

[130] J.R. Oppenheimer, *Science and the Common Understanding.,* Oxford University Press: New York.

[131] N. Bohr, *Atomic Physics and Human Knowledge.,* John Wiley & Sons: New York.

[132] V.A. Lektorskii, *Classical and Nonclassical Epistemology.,* Editorial URSS: Moscow. [in Russian]

[133] I.T. Kasavin, The truth as a norm. The truth as a description. Truth, description, and expertise.*New Philosophical Encyclopedia.,* Mysl: Moscow. [in Russian]

[134] L.A. Mikeshina, Epistemology and cognitive science: basic categories and principles of interaction.*Cognitive Approach.,* Philosophy, Cognitive Science, and Cognitive Disciplines: Moscow, . [in Russian]

[135] *Asvaghosa, The Awakening of Faith.,* Open Court: Chicago, Illinois, .D.T. Suzuki, Ed.

[136] V. Boss, *Lectures on Mathematics*, Moscow: Editorial URSS, 2004–2011, vol. 8, *The Theory of Groups*, 2007 [in Russian].

[137] Mathematics, In: *Large Encyclopedic Dictionary*, Moscow: Bol'shaya Rossiiskaya entsiklopediya, 2000 [in Russian].

[138] A. Watts, *The Way of Zen.,* Pantheon: New York.

[139] S. Radhakrishnan, *Indian Philosophy.,* Oxford University Press.

[140] A.N. Leont'ev, *Activity, Consciousness, and Personality.,* Prentice-Hall.

[141] V.F. Petrenko, and V.V. Kucherenko, "Meditation as a form of non-mediated cognition", *Vopr. Filos,* vol. 83, no. 8.

[142] G.W. Leibniz, *Collected Works in Four Volumes*Mysl: Moscow.

[143] P.A. Schilpp, *Albert Einstein. Philosopher-Scientist.,* Library of Living Philosophers: Evanston.

[144] H. Jiyu-Kennett, *Selling Water by the River: A Manual of Zen Training.,* Pantheon Books.

[145] A. Einstein, "Strahlungs-emission und -absorption nach der Quantentheorie", *Verh. Dtsch. Phys. Ges.,* vol. 18, p. 318.

[146] A. Aspect, P. Grangier, and G. Roger, "Experimental realization of Einstein–Podolsky–Rosen–Bohm Gedanken experiment: A new violation of Bell's inequality", *Phys. Rev. Lett.,* vol. 49, no. 2, pp. 91-94.[http://dx.doi.org/10.1103/PhysRevLett.49.91]

[147] E. Schrödinger, *Mind and Matter.,* University Press: Cambridge.

[148] E. Wigner, *Symmetries and Reflections: Scientific Essays.,* Indiana University Press: Bloomington, .

[149] M. Proietti, A. Pickston, F. Graffitti, P. Barrow, D. Kundys, C. Branciard, M. Ringbauer, and A. Fedrizzi, "Experimental test of local observer independence", *Sci. Adv.,* vol. 5, no. 9, p. eaaw9832.[http://dx.doi.org/10.1126/sciadv.aaw9832] [PMID: 31555731]

[150] Vasubandhu, *Verses on the Treasury of Abhidharma*, 1994 [in Russian].

[151] S.L. Rubinshtein, *Man and the World.,* Moscow, . [in Russian]

[152] A. Einstein, *Ideas and Options.,* Crown Publishers: New York.

[153] V.F. Petrenko, *Foundations of Psychosemantics.*, Eksmos: Moscow, . [in Russian]

[154] L.S. Vygotsky, Collected Works

[155] H. Poincare, L'hypothese des quanta, *Revue scientifique*, 4 (17), 225, 1912.

[156] S. Chatterjee, and D. Datta, *An Introduction to Indian Philosophy.*, Rupa Publications.

[157] N.V. Abaev, *Psychological Aspects of Buddhism.*, Nauka: Novosibirsk, . [in Russian]

[158] *Anthology of Taoist Philosophy (composed by V.V. Malyavin and B.B. Vinogrodskii).*, Komarov I K: Moscow, . [in Russian]

[159] F.I. Shcherbatskoi, *Selected Works on Buddhism.*, Nauka: Moscow, . [in Russian]

[160] H.P. Stapp, "S-Matrix interpretation of quantum theory", *Phys. Rev. D Part. Fields,* vol. 3, no. 6, pp. 1303-1320.[http://dx.doi.org/10.1103/PhysRevD.3.1303]

[161] G.H. Mullin, *Readings on the Six Yogas of Naropa.*, Snow Lion Publications: Ithaca, NY, .

[162] R. Moody, *Reflections on Life after Life.*, Stackpole Books: Harrisburg, .

[163] S. Rinpoche, *The Tibetan Book of Living and Dying.*, Harper Collins Publisher Inc.: San Francisco, .

[164] A.P. Suprun, Relativist psychology: a new concept of psychological measurement[http://dx.doi.org/10.11621/pir.2009.0013]

[165] K. Thorne, *Black Holes and Time Wraps: Einstein's Outrageous Legacy,* .

[166] R.P. Kerr, "Gravitational field of a spinning mass as an example of algebraically special metrics", *Phys. Rev. Lett.,* vol. 11, no. 5, pp. 237-238.[http://dx.doi.org/10.1103/PhysRevLett.11.237]

[167] G.C. Debney, R.P. Kerr, and A. Schild, "Solutions of the Einstein and Einstein□Maxwell Equations", *J. Math. Phys.,* vol. 10, no. 10, pp. 1842-1854.[http://dx.doi.org/10.1063/1.1664769]

[168] R.A. Moody, *Life after Life.*, Mockingbird Books, .

[169] A.J.S. Hamilton, and P.P. Avelino, "The physics of the relativistic counter-streaming instability that drives mass inflation inside black holes", *Phys. Rep.,* vol. 495, no. 1, pp. 1-32.[http://dx.doi.org/10.1016/j.physrep.2010.06.002]

[170] A.W. Wiggins, and C.M. Wynn, *The Five Biggest Unsolved Problems in Science.*, John Wiley & Sons, Inc., .

[171] W. Hampson, *Paradoxes of Nature and Science.*, Cassell and Co., Ltd.: London, .

[172] J. Brockman, Ed., *This Idea Must Die: Scientific Theories That Are Blocking Progress.*, Edge Foundation, Inc., .

[173] I.S Helland, "Conceptual variables, quantum theory, and statistical inference theory",

[174] L. Smolin, *The Trouble with Physics: The Rise of String Theory, the Fall of a Science, and What Comes Next.*, Houghton Mifflin Company: Boston, New York, .

SUBJECT INDEX

www.ingramcontent.com/pod-product-compliance
Lightning Source LLC
Chambersburg PA
CBHW050841220326
41598CB00006B/421